VOLUME V

Introduction to Microelectronic Fabrication

MODULAR SERIES
ON SOLID STATE DEVICES
Gerold W. Neudeck and Robert F. Pierret, Editors

VOLUME V
Introduction to Microelectronic Fabrication

RICHARD C. JAEGER

Auburn University

ADDISON-WESLEY PUBLISHING COMPANY

READING, MASSACHUSETTS

MENLO PARK, CALIFORNIA • NEW YORK • DON MILLS, ONTARIO
WOKINGHAM, ENGLAND • AMSTERDAM • BONN • SYDNEY
SINGAPORE • TOKYO • MADRID • BOGOTÁ
SANTIAGO • SAN JUAN

This book is in the
Addison-Wesley Modular Series on Solid State Devices

Library of Congress Cataloging-in-Publication Data

Jaeger, Richard C.
  Introduction to microelectronic fabrication.

  (Modular series on solid state devices; v. 5)
  Includes index.
  1. Integrated circuits — Very large scale integration —
Design and construction — Congresses.
I. Title. II. Series
TK7874.J33 1988 621.381'73 86-22128
ISBN 0-201-14695-9

Reprinted with corrections May, 1993.

19 20 -MA- 00 99

To My Family — Joan, Peter, and Stephanie

Foreword

The spectacular advance in the development and application of integrated-circuit technology is unparalleled in our history, and the rapid growth of the IC industry has led to the emergence of microelectronics process engineering as a new discipline. The pervasive use of integrated circuits also requires that a wide range of engineers in the highly competitive electronics industry have an understanding of the design and limitations of integrated circuits. This text presents an introduction to the basic processes common to all IC technologies and provides a base for understanding more advanced processing and what can and cannot be achieved through integrated-circuit fabrication.

The book has evolved from notes developed over the past seven years for a course which introduces seniors and beginning graduate students to the fabrication of solid state devices and integrated circuits. It assumes a basic knowledge of the material properties of silicon, and we use Volume I of this Series as a companion text in the fabrication course. This work also assumes a minimal knowledge of the existence of integrated circuits and of the terminal behavior of electronic components such as resistors, diodes, and MOS and bipolar transistors. In order to contain the scope of the material, we deal only with basic silicon processing.

In our curriculum, other courses cover solid state materials, unipolar and bipolar device physics, and digital and analog integrated-circuit design. The material and device courses use Volumes 1–4 of this Series, and every attempt has been made to make the notation coincide with the earlier volumes, although some differences may have slipped through.

The goal of this book is to present the basic information necessary to prepare the student for more advanced processing and design courses. The special problems of VLSI fabrication are left to advanced processing texts, although a number of problem areas are mentioned throughout the book. Chapters 2–5 focus on the basic processes used in fabrication, including lithography, oxidation, diffusion, ion implantation, and thin-film deposition. Interconnection technology, packaging, and yield, often neglected, are covered in Chapters 7 and 8. It is my belief that the student must also understand the basic interaction between process design, device design, and device layout. For this reason, Chapters 9 and 10 on MOS and bipolar process integration have been included in this book.

The material of the book is designed to be covered in one semester. On the quarter system at Auburn, we cover Chapters 1–5 and most of the material from Chapters 9 and 10. The problems have been developed to both reinforce and extend the material in the chapters. In our case, the microelectronics fabrication course is also accompanied by a laboratory which strongly reinforces the classroom material. The test chip of Fig. 2.7 is fabricated and tested during this laboratory, and a few of the problems are related to characterization of structures on this chip. We can provide pattern generator data for this mask set or can supply wafers fabricated by previous classes on an "as-available" basis.

I must also recognize a number of previous books which have obviously influenced the preparation of this text. These include *The Theory and Practice of Microelectronics* and *VLSI Fabrication Principles* by S. K. Ghandhi, *Basic Integrated Circuit Engineering* by D. J. Hamilton and W. G. Howard, *Integrated Circuit Engineering* by A. H. Glaser and G. E. Subak-Sharpe, *Microelectronics Processing and Device Design* by R. A. Colclaser, and *Semiconductor Devices — Physics and Technology* by S. M. Sze.

I thank my family for putting up with the countless hours of work which have gone into the preparation of this book. In particular, I want to thank my wife Joan for the many hours that she spent in the library tracking down and verifying errant references. During preparation of this text, we found that the references in many popular, recently published books were virtually useless because they contained so many errors. We have done our best to ensure that this is not the case with this text.

Thanks also must go to my colleagues who have helped with this book, especially to Jim L. Davidson for his suggestions, and to our laboratory managers, Walter Power and Charles Ellis, who have been instrumental in developing the processes for the microelectronics class chips.

Richard C. Jaeger

Contents

3   Thermal Oxidation of Silicon

4   Diffusion

5   Ion Implantation

6   Film Deposition

7   Interconnections and Contacts

8   Packaging and Yield

9 MOS Process Integration

10 Bipolar Process Integration

1 / An Overview of Microelectronic Fabrication

1.1 A HISTORICAL PERSPECTIVE

In this volume we will develop an understanding of the basic processes used in monolithic integrated-circuit fabrication. Silicon is the dominant material used throughout the integrated-circuit industry today, and in order to conserve space only silicon processing will be discussed in this book. However, all of the basic processes discussed here are applicable to the fabrication of gallium arsenide integrated circuits (ICs) and thick- and thin-film hybrid ICs.

Germanium was one of the first materials to receive wide attention for use in semiconductor device fabrication, but it was rapidly replaced by silicon during the early 1960s. Silicon emerged as the dominant material because it was found to have major processing advantages. Silicon can easily be oxidized to form silicon dioxide. Silicon dioxide was found to be not only a high-quality insulator but also an excellent barrier layer for the selective diffusion steps needed in integrated-circuit fabrication.

Silicon was also shown to have a number of ancillary advantages. It is a very abundant element in nature, providing the possibility of a low-cost starting material. It has a wider bandgap than germanium and can therefore operate at higher temperatures than germanium. In retrospect it appears that the processing advantages were the dominant reasons for the emergence of silicon over other semiconductor materials.

The first successful fabrication techniques produced single transistors on a silicon die 1 to 2 mm on a side. Early integrated circuits, fabricated at Texas Instruments and Fairchild Semiconductor, included several transistors and resistors to make simple logic gates and amplifier circuits. From this modest beginning, we have reached integration levels of several million components on a 7 mm × 7 mm die. For example, a one-megabit dynamic random-access memory (DRAM) chip has more than 1,000,000 transistors and more than 1,000,000 capacitors in the memory array, as well as tens of thousands of transistors in the access and decoding circuits. The level of integration has been doubling every one to two years since the early 1960s.

1

One-megabit RAMs are currently being produced with photolithographic features measuring between 1 and 2 microns (μm). MOS transistors with dimensions approximately ten times smaller (0.1 to 0.2 μm) have already been fabricated in research laboratories. So we still have at least a factor of 100 to go in terms of integrated-circuit density, provided manufacturable fabrication processes can be developed for these sub-micron dimensions.

Early fabrication used silicon wafers which had 1- and then 2-in. diameters. The size of the wafers has steadily increased to the point where 4-, 5-, and 6-in. wafers are now in production. Wafers with 8-in. diameters have been successfully produced by silicon wafer manufacturers.

The larger the diameter of the wafer, the more integrated-circuit dice can be produced at one time. Many wafers are processed at the same time. The same silicon chip is replicated as many times as possible on a silicon wafer of a given size. Figure 1.1 shows the approximate number of 5 $\times$ 5 mm dice that fit on a wafer of a given diameter. Processing costs per wafer are relatively independent of wafer size, so the cost per die is lower for larger wafer sizes. Thus there are strong economic forces driving the integrated-circuit industry to continually move to larger and larger wafer sizes.

Here we see a problem with the units of measure used to describe integrated circuits. Horizontal dimensions were originally specified in mils (1 mil = 0.001 in.), whereas specification of the shallower vertical dimensions commonly made use of the metric system. Today, most of the dimensions are specified using the metric system, although English units are occasionally still used. Throughout the rest of this book, we will attempt to make consistent use of metric units.

1.2   AN OVERVIEW OF MONOLITHIC FABRICATION PROCESSES AND STRUCTURES

Monolithic integrated-circuit fabrication can be illustrated by studying the basic cross sections of MOS and bipolar transistors in Figs. 1.2 and 1.3. The n-channel MOS transistor is formed in a p-type substrate. Source/drain regions are formed by selectively converting shallow regions at the surface to n-type material. Thin and thick silicon dioxide regions on the surface form the gate insulator of the transistor and serve to isolate one device from another. A thin film of polysilicon is used to form the gate of the transistor, and aluminum is used to make contact to the source and drain. Interconnections between devices can be made using the diffusions and the layers of polysilicon and metal.

The bipolar transistor has alternating n- and p-type regions selectively fabricated on a p-type substrate. Silicon dioxide is again used as an insulator, and a metal such as aluminum is used to make electrical contact to the emitter, base, and collector of the transistor.

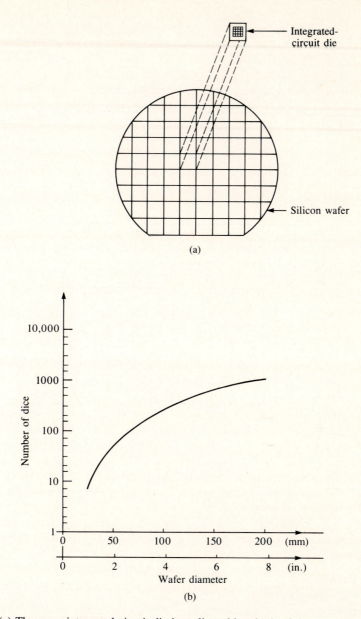

Fig. 1.1   (a) The same integrated-circuit die is replicated hundreds of times on a typical silicon wafer; (b) the graph gives the approximate number of 5 × 5 mm dice which can be fabricated on wafers of different diameters.

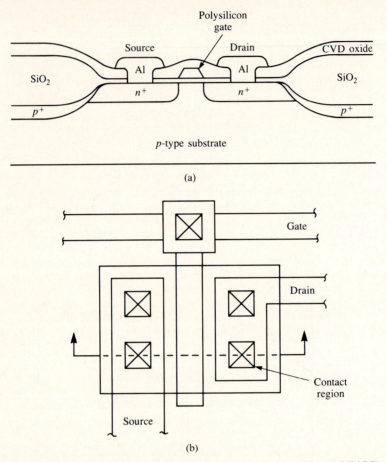

Fig. 1.2 The basic structure of an *n*-channel metal-oxide-semiconductor (NMOS) transistor structure. (a) The vertical cross section through the transistor; (b) a composite top view of the masks used to fabricate the transistor in (a).

These structures are fabricated through the repeated application of a number of basic processing steps:

- Oxidation
- Photolithography
- Etching
- Diffusion
- Evaporation or sputtering
- Chemical vapor deposition (CVD)
- Ion implantation
- Epitaxy

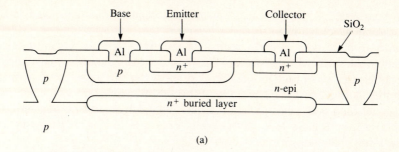

(a)

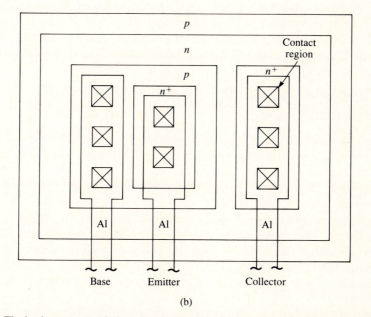

(b)

Fig. 1.3   The basic structure of a junction-isolated bipolar transistor. (a) The vertical cross section through the transistor; (b) a composite top view of the masks used to fabricate the transistor in (a).

Silicon dioxide can be formed by heating a silicon wafer to a high temperature (1000 to 1200 °C) in the presence of oxygen. This process is called *oxidation*. Metal films can be deposited through evaporation by heating the metal to its melting point in a vacuum. Thin films of silicon nitride, silicon dioxide, and polysilicon can all be formed through a process known as *chemical vapor deposition* (CVD), in which the material is deposited out of a gaseous mixture onto the surface of the wafer. Metals and insulators may also be deposited by a process called *sputtering*.

Shallow *n*- and *p*-type layers are formed by high-temperature (1000 to 1200 °C) *diffusion* of donor or acceptor impurities into silicon or by *ion implantation*, in which the wafer is bombarded with high-energy donor or acceptor atoms generated in a high-voltage particle accelerator.

In order to build devices and circuits, the n- and p-type regions must be formed selectively in the surface of the wafer. Silicon dioxide, silicon nitride, polysilicon, and other materials can all be used to mask areas of the wafer surface to prevent penetration of impurities during ion implantation or diffusion. Windows are cut in the masking material by etching with acids or in a plasma. Window patterns are transferred to the wafer surface from a mask through the use of optical techniques. The masks are also produced using photographic reduction techniques.

Photolithography includes the overall process of mask fabrication as well as the process of transferring patterns from the masks to the surface of the wafer. The photolithographic process is critical to the production of integrated circuits, and the number of mask steps is often used as a measure of complexity when comparing fabrication processes.

1.3 METAL-OXIDE-SEMICONDUCTOR (MOS) PROCESSES

1.3.1 Basic NMOS Process

A possible process flow for a basic n-channel MOS process (NMOS) is shown in Figs. 1.4 and 1.5. The starting wafer is first oxidized to form a thin-pad oxide layer of silicon dioxide (SiO_2) which protects the silicon surface. Silicon nitride is then deposited by a low-pressure chemical vapor deposition (LPCVD) process. Mask #1 defines the active transistor areas. The nitride/oxide sandwich is etched away everywhere except where transistors are to be formed. A boron implantation is performed and followed by an oxidation step. The nitride serves as both an implantation mask and an oxidation mask. After the nitride and thin oxide padding layers are removed, a new thin layer of oxide is grown to serve as the gate oxide for the MOS transistors. Following gate-oxide growth, a boron implantation is commonly used to adjust the threshold voltage to the desired value.

Polysilicon is deposited over the complete wafer using a CVD process. The second mask defines the polysilicon gate region of the transistors. Polysilicon is etched away everywhere except over the gate regions and the areas used for interconnection. Next, the source/drain regions are implanted through the thin oxide regions. The implanted impurity may be driven in deeper with a high-temperature diffusion step. More oxide is deposited on the surface, and contact openings are defined by the third mask step. Metal is deposited over the wafer surface by evaporation or sputtering. The fourth mask step is used to define the interconnection pattern which will be etched in the metal. A passivation layer of phosphosilicate glass (not shown in Fig. 1.4) is deposited on the wafer surface, and the final mask (#5) is used to define windows so that bonding wires can be attached to pads on the periphery of the IC die.

This simple process requires five mask steps. Note that these mask steps use subtractive processes. The entire surface of the wafer is first coated with a desired material, and then most of the material is removed by wet chemical or plasma etching.

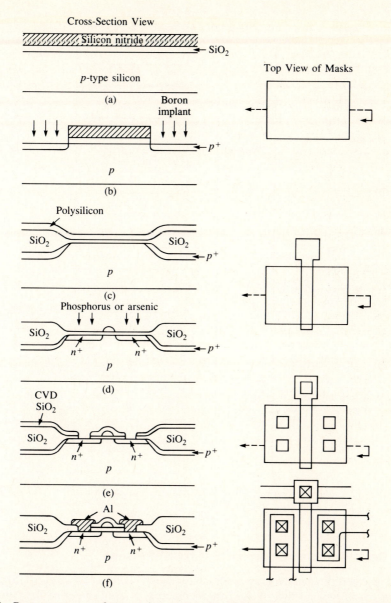

Fig. 1.4 Process sequence for a semirecessed oxide NMOS process. (a) Silicon wafer covered with silicon nitride over a thin padding layer of silicon dioxide; (b) etched wafer after first mask step. A boron implant is used to help control field oxide threshold; (c) structure following nitride removal and polysilicon deposition; (d) wafer after second mask step and etching of polysilicon; (e) the third mask has been used to open contact windows following silicon dioxide deposition; (f) final structure following metal deposition and patterning with fourth mask.

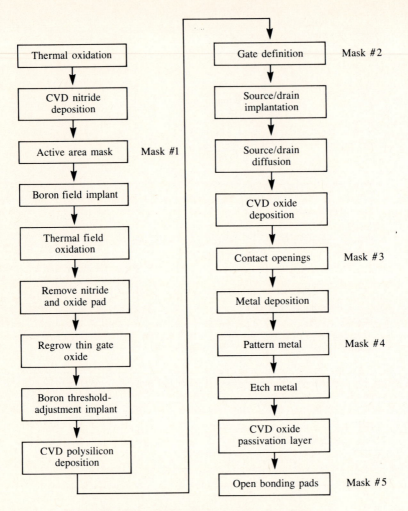

Fig. 1.5 Basic NMOS process flowchart.

1.3.2 Basic Complementary MOS (CMOS) Process

Figure 1.6 shows the mask sequence for a basic complementary MOS (CMOS) process. One new mask, beyond that of the NMOS process, is used to define the "*p*-well" or "*p*-tub," which serves as the substrate for the *n*-channel devices. A second new mask step is used to define the source/drain regions of the *p*-channel transistors. Additional masks may be used to adjust the threshold voltage of the MOS transistors and are very common in state-of-the-art NMOS and CMOS processes.

Some recent CMOS processes use an *n*-well instead of a *p*-well. The *n*-well can be added to an existing NMOS process with a minimum of change, and it permits high-

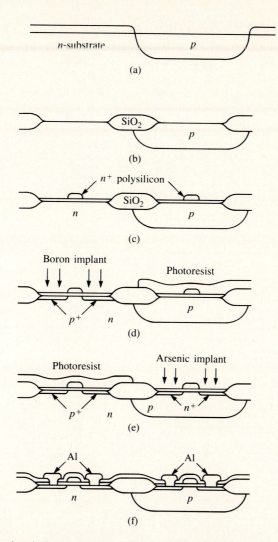

Fig. 1.6 Cross-sectional views at major steps in a basic CMOS process. (a) Following p-well diffusion, (b) following selective oxidation, and (c) following gate oxidation and polysilicon gate definition; (d) PMOS source/drain implantation; (e) NMOS source/drain implantation; (f) structure following contact and metal mask steps.

performance NMOS and CMOS on the same chip. Twin-well processes have also been developed recently. Both a p-well and an n-well are formed in a lightly doped substrate, and the n- and p-channel devices can each be optimized for highest performance. Twin-well very large-scale integration (VLSI) processes use lightly doped layers grown on heavily doped substrates to suppress a CMOS failure mode called *latchup*.

1.4   BASIC BIPOLAR PROCESSING

Basic bipolar fabrication is somewhat more complex than single-channel MOS process-ing, as indicated in Figs. 1.7 and 1.8. A *p*-type silicon wafer is oxidized, and the first

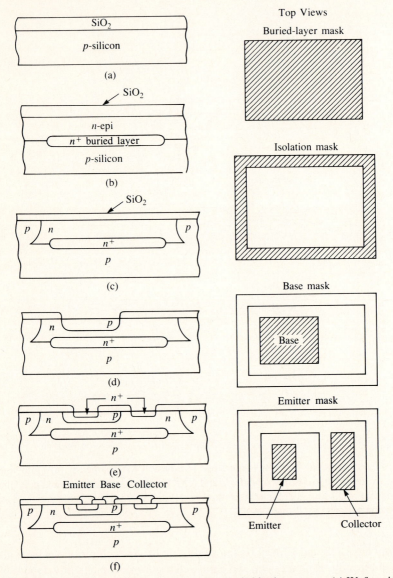

Fig. 1.7   Cross-sectional view of the major steps in a basic bipolar process. (a) Wafer with silicon dioxide layer; (b) following buried-layer diffusion using first mask, and subsequent epitaxial layer growth and oxidation; (c) following deep-isolation diffusion using second mask; (d) following boron-base diffusion using third mask; (e) fourth mask defines emitter and collector contact regions; (f) final structure following contact and metal mask steps.

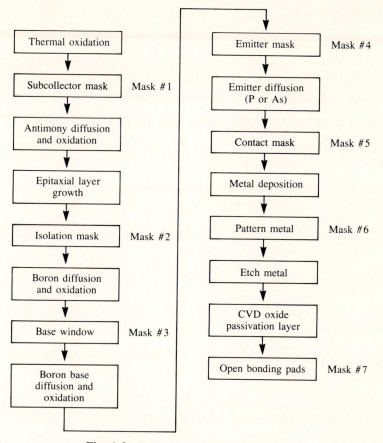

Fig. 1.8 Basic bipolar process flowchart.

mask is used to define a diffused region called the *buried layer* or *subcollector*. This diffusion is used to reduce the collector resistance of the bipolar transistor. Following the buried-layer diffusion, a process called *epitaxy* is used to grow single-crystal n-type silicon on top of the silicon wafer. The epitaxial growth process results in a high-quality silicon layer with the same crystal structure as the original silicon wafer. An oxide layer is then grown on the wafer. Mask two is used to open windows for a deep p-diffusion, which is used to isolate one bipolar transistor from another. Another oxidation follows the isolation diffusion. Mask three opens windows in the oxide for the p-type base diffusion. The wafer is usually oxidized during the base diffusion, and mask four is used to open windows for the emitter diffusion. The same diffusion step places an n^+ region under the collector contact to ensure that a good ohmic contact will be formed during subsequent metallization. Masks five, six, and seven are used to open contact windows, pattern the metallization layer, and open windows in the passivation layer just as in the NMOS process described in Section 1.3. Thus the basic bipolar process requires seven mask levels compared with five for the basic NMOS process.

After the MOS or bipolar process is completed, each die on the wafer is tested, and bad dice are marked with ink. The wafer is then sawed apart. Good dice are mounted in various packages for final testing and subsequent sale or use.

The rest of this book concentrates on the basic processes used in the fabrication of monolithic integrated circuits. Chapters 2 through 8 discuss mask making and pattern definition, oxidation, diffusion, ion implantation, film deposition, interconnections and contacts, and packaging and yield. The last two chapters introduce the integration of process, layout, and device design for MOS and bipolar technologies.

PROBLEMS

1.1 The curve in Fig. 1.1b represents the approximate number of chips on a wafer of a given diameter. Determine the exact number of 5×5 mm dice that will fit on a wafer with a diameter of 100 mm. (The number indicated on the curve is 254.)

1.2 The cost of processing a wafer in a particular process is $400. Assume that 35% of the fabricated dice are good. Find the number of dice, using Fig. 1.1b.

(a) Determine the cost per good die for a 75-mm wafer.

(b) Repeat for a 150-mm wafer.

1.3 A certain silicon-gate NMOS transistor occupies an area of $25 \lambda^2$ where λ is the minimum lithographic feature size.

(a) How many MOS transistors can fit on a 5×5 mm die if $\lambda = 10 \ \mu m$?

(b) $2.5 \ \mu m$?

(c) $1 \ \mu m$?

1.4 A simple *pn* junction diode is shown in cross section in Fig. P1.4. Make a possible process flowchart for fabrication of this structure, including mask steps.

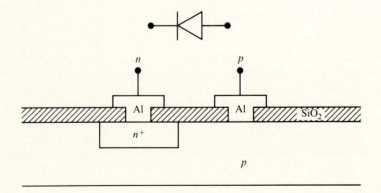

Fig. P1.4

1.5 Draw a set of contact and metal masks for the bipolar transistor of Fig. 1.7. Use square contact windows with one contact to the emitter and two contacts to the base and collector regions.

2 / Lithography

In order to produce an integrated circuit, thin films of various materials are used as barriers to the diffusion or implantation of impurity atoms, or as insulators between conductive materials and the silicon substrate. Holes or windows are cut through this barrier material wherever impurity penetration or contact is desired.

Masks contain the patterns of windows which are transferred to the surface of the silicon wafer using a process called *photolithography*. Photolithography makes use of a highly refined version of the photoengraving process. The patterns are first transferred from the mask to a light-sensitive material called *photoresist*. Chemical or plasma etching is then used to transfer the pattern from the photoresist to the barrier material on the surface of the wafer. Each mask step requires successful completion of numerous processing steps, and the complexity of an integrated-circuit process is often measured by the number of photographic masks used during fabrication. This chapter will explore the lithographic process, including mask fabrication, photoresist processes, and etching.

2.1 THE PHOTOLITHOGRAPHIC PROCESS

Photolithography encompasses all the steps involved in transferring a pattern from a mask to the surface of the silicon wafer. The various steps of the basic photolithographic process given in Figs. 2.1 and 2.2 will each be discussed in detail below.

Ultraclean conditions must be maintained during the lithography process. Any dust particles on the original substrate or that fall on the substrate during processing can result in defects in the final resist coating. Even if defects occur in only 10% of the chip sites at each mask step, less than 50% of the chips will be functional after a seven-mask process is completed. Vertical laminar-flow hoods in clean rooms are used to prevent particulate contamination throughout the fabrication process. Clean rooms use filtration to remove particles from the air and are rated by the maximum number of particles per cubic foot or cubic meter of air, as shown in Table 2.1. Clean rooms have evolved from the Class 10,000 to the Class 10 and even Class 1 facilities now being used for VLSI processing. For comparison, each cubic foot of ordinary room air has several million dust particles exceeding a size of 0.5 μm.

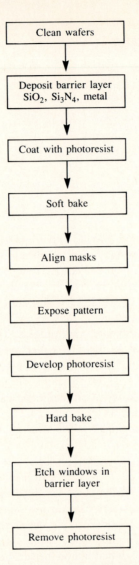

Fig. 2.1 Steps of the photolithographic process.

2.1.1 Wafer Cleaning

Prior to use, wafers are chemically cleaned to remove particulate matter on the surface as well as any traces of organic, ionic, and metallic impurities. A cleaning step in a solution of hydrofluoric acid is used to remove any oxide which may have formed on the wafer surface. A typical cleaning process is presented in Table 2.2.

One very important chemical used in wafer cleaning and throughout microelectronic fabrication processes is deionized (DI) water. DI water is highly purified and filtered to

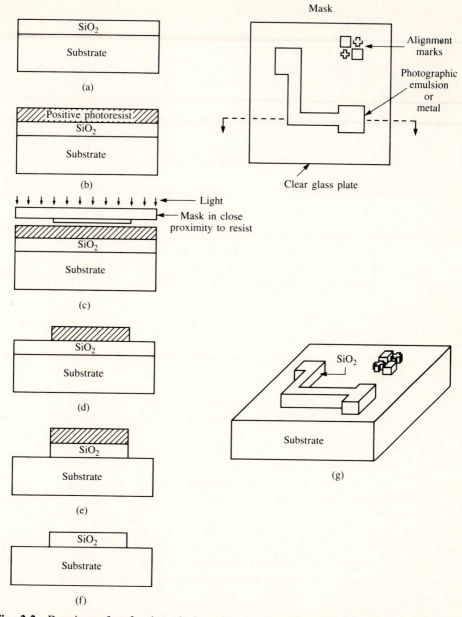

Fig. 2.2 Drawings of wafer through the various steps of the photolithographic process. (a) Substrate covered with silicon dioxide barrier layer; (b) positive photoresist applied to the surface of the wafer; (c) mask in close proximity to the surface of the resist-covered wafer; (d) substrate following resist exposure and development; (e) substrate following etching of the silicon dioxide layer; (f) oxide barrier on wafer surface after resist removal; (g) view of substrate with silicon dioxide pattern on the surface.

Table 2.1 Ratings by Class of Effectiveness of Filtration in Clean Rooms.

Class	Number of 0.5-μm particles per ft^3 (m^3)		Number of 5-μm particles per ft^3 (m^3)	
10,000	10,000	(350,000)	65	(23,000)
1,000	1,000	(35,000)	6.5	(2,300)*
100	100	(3,500)	0.65	(230)*
10	10	(350)	0.065	(23)*
1	1	(35)*	0.0065	(2.3)*

*It is very difficult to measure particulate counts below 10 per ft^3.

Table 2.2 Silicon Wafer Cleaning Procedure.[4,5]

A. Solvent Removal
1. Immerse in boiling trichloroethylene (TCE) for 3 min.
2. Immerse in boiling acetone for 3 min.
3. Immerse in boiling methyl alcohol for 3 min.
4. Wash in DI water for 3 min.

B. Removal of Residual Organic/Ionic Contamination
1. Immerse in a (5:1:1) solution of H_2O–NH_4OH–H_2O_2; heat solution to 75–80 °C and hold for 10 min.
2. Quench the solution under running DI water for 1 min.
3. Wash in DI water for 5 min.

C. Hydrous Oxide Removal
1. Immerse in a (1:50) solution of HF–H_2O for 15 sec.
2. Wash in running DI water with agitation for 30 sec.

D. Heavy Metal Clean
1. Immerse in a (6:1:1) solution of H_2O–HCl–H_2O_2 for 10 min at a temperature of 75–80 °C.
2. Quench the solution under running DI water for 1 min.
3. Wash in running DI water for 20 min.

remove all traces of ionic, particulate, and bacterial contamination. The theoretical resistivity of pure water at 25 °C is 18.3 Mohm-cm. Typical DI water systems achieve resistivities of 18 Mohm-cm with fewer than 1.2 colonies of bacteria per milliliter and with no particles larger than 0.25 μm.

2.1.2 Barrier Layer Formation

After cleaning, the silicon wafer is covered with the material which will serve as a barrier layer. The most common material is silicon dioxide (SiO_2), so we will use it as an example here. Silicon nitride (Si_3N_4), polysilicon, photoresist, and metals are also routinely used as barrier materials at different points in a given process flow. Later chapters will discuss thermal oxidation, chemical vapor deposition, sputtering, and vacuum evaporation processes, all of which are used to produce thin layers of these materials.

The original silicon wafer has a metallic gray appearance. Once a silicon dioxide layer is formed on the silicon wafer, the surface will have a color which depends on the SiO_2 thickness. The finished wafer will have regions with many different thicknesses. Each region will produce a different color, resulting in beautiful, multicolored IC images, photographs of which appear in many books and magazines.

2.1.3   Photoresist Application

After formation of the SiO_2 layer, the surface of the wafer is coated with a light-sensitive material called *photoresist*. The surface must be clean and dry to ensure good photoresist adhesion. Freshly oxidized wafers may be directly coated, but if the wafers have been stored, they should be carefully cleaned and dried prior to application of the resist. A liquid adhesion promoter is often applied just prior to resist application.

Photoresist is typically applied in liquid form. The wafer is held on a vacuum chuck and then spun at high speed for 30 to 60 sec to produce a thin uniform layer. Speeds of 1000 to 5000 rpm result in layers ranging from 2.5 to 0.5 μm, respectively. The actual thickness of the resist depends on its viscosity and is inversely proportional to the square root of the spinning speed.

2.1.4   Soft Baking

A drying step called *soft baking* or *prebaking* is used to improve adhesion and remove solvent from the photoresist. Times range from 10 to 30 min in an oven at 80 to 90 °C in an air or nitrogen atmosphere. The soft-baking process is specified on the resist manufacturer's data sheet and should be followed closely. After soft baking, the photoresist is ready for mask alignment and exposure.

2.1.5   Mask Alignment

A photomask, a square glass plate with a patterned emulsion or metal film on one side, is placed over the wafer. Each mask following the first must be carefully aligned to the previous pattern on the wafer. Much of the alignment has traditionally involved manual operation of alignment equipment. VLSI designs with minimum-size geometrical features measuring 1.25 μm (minimum linewidth or space) require an alignment tolerance of better than ± 0.25 μm. Computer-controlled alignment equipment has been developed to achieve this level of alignment precision.

With manual alignment equipment, the wafer is held on a vacuum chuck and carefully moved into position below the mask using an adjustable *x-y* stage. The mask is spaced 25 to 125 μm above the surface of the wafer during alignment. If contact printing is being used, the mask is brought into contact with the wafer after alignment.

Alignment marks are introduced on each mask and transferred to the wafer as part of the integrated-circuit pattern. The marks are used to align each new mask level to one of the previous levels. A sample set of alignment marks is shown in Fig. 2.3. For certain mask levels, the cross on the mask is placed in a box on the wafer. For other mask levels, the box on the mask is placed over a cross on the wafer. The choice depends on the type of resist used during a given photolithographic step. Split-field optics are used to simultaneously align two well-separated areas of the wafer.

2.1.6 Photoresist Exposure and Development

Following alignment, the photoresist is exposed through the mask with high-intensity ultraviolet light. Resist is exposed wherever silicon dioxide is to be removed. The photoresist is developed with a process very similar to that used for developing ordinary photographic film, using a developer supplied by the photoresist manufacturer. Any resist which has been exposed to ultraviolet light is washed away, leaving bare silicon dioxide in the exposed areas of Fig. 2.2d. A photoresist acting in the manner just described is called a *positive resist,* and the mask contains a copy of the pattern which will remain on the surface of the wafer. Windows are opened wherever the exposing light passes through the mask.

Negative photoresists can also be used. A negative resist remains on the surface wherever it is exposed. Figure 2.4 shows simple examples of the patterns transferred to a silicon dioxide barrier layer using positive and negative photoresists with the same mask. Negative resists were widely used in early integrated-circuit processing. However, positive resist yields better process control in small-geometry structures and is now the main type of resist used in VLSI processes.

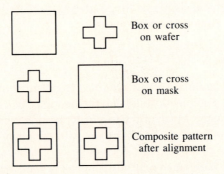

Fig. 2.3 A simple set of alignment marks. At some steps a cross may be aligned within a box. At others, a box may be placed around the cross. The choice depends on the type of resist being used at a given mask step.

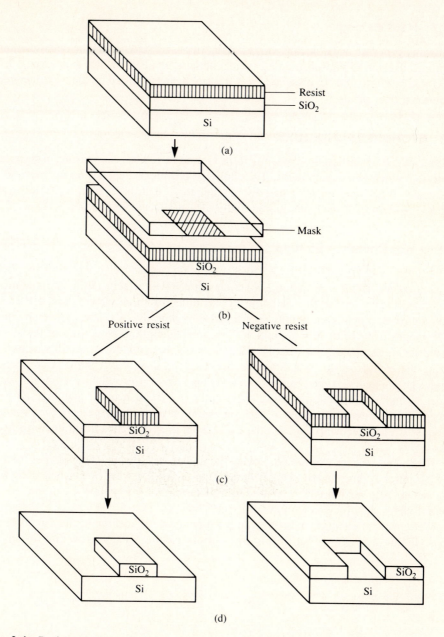

Fig. 2.4 Resist and silicon dioxide patterns following photolithography with positive and negative resists.

2.1.7 Hard Baking

Following exposure and development, a baking step is used to harden the photoresist and improve adhesion to the substrate. A typical process involves baking in an oven for 20 to 30 min at 120 to 180 °C. Details of this step are again specified on the manufacturer's photoresist data sheets.

2.2 ETCHING TECHNIQUES

Chemical etching in liquid or gaseous form is used to remove any barrier material not protected by hardened photoresist. The choice of chemicals depends on the material to be etched. A high degree of selectivity is required so that the etchant will remove the unprotected barrier layer much more rapidly than it attacks the photoresist layer.

2.2.1 Wet Chemical Etching

A buffered oxide etch (BOE or BHF) is commonly used to etch windows in silicon dioxide layers. BOE is a solution containing hydrofluoric acid (HF), and etching is performed by immersing the wafers in the solution. At room temperature, HF etches silicon dioxide much more rapidly than it etches photoresist or silicon. The etch rate in BOE ranges from 10 to 100 nm/min at 25 °C, depending on the density of the silicon dioxide film. Etch rate is temperature-dependent, and temperature is carefully monitored during the etch process. In addition, etch rates depend on the type of oxide present. Oxides grown in dry oxygen etch more slowly than those grown in the presence of water vapor. A high concentration of phosphorus in the oxide enhances the etch rate, whereas a reduced etch rate occurs when a high concentration of boron is present. High concentrations of these elements convert the SiO_2 layer to a phosphosilicate or borosilicate glass.

HF and water both wet silicon dioxide but do not wet silicon. The length of the etch process may be controlled by visually monitoring test wafers which are etched along with the actual integrated-circuit wafers. Occurrence of a hydrophobic condition on the control wafer signals completion of the etch step.

Wet chemical etching tends to be an isotropic process, etching equally in all directions. Figure 2.5a shows the result of isotropic etching of a narrow line in silicon

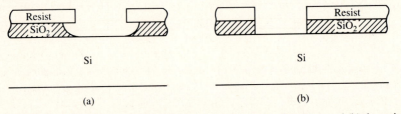

Fig. 2.5 Etching profiles obtained with (a) isotropic wet chemical etching and (b) dry anisotropic etching in a plasma or reactive-ion etching system.

dioxide. The etching process has etched under the resist by a distance equal to the thickness of the film. This "etch bias" becomes a serious problem in processes requiring linewidths with dimensions similar to the thickness of the film.

2.2.2 Dry Etching

Dry etching processes are widely used in VLSI fabrication. Highly anisotropic etching profiles can be obtained as shown in Fig. 2.5b, avoiding the undercutting problem of Fig. 2.5a characteristic of wet processes. Dry processes require only small amounts of reactant gases, whereas wet etching requires disposal of relatively large amounts of liquid chemical wastes.

Plasma etching immerses the wafers in a gaseous plasma created by RF excitation in a vacuum system. The plasma contains fluorine or chlorine ions which etch silicon dioxide. The RF power source typically operates at a frequency of 13.56 MHz, which is set aside by the Federal Communications Commission for industrial and scientific purposes.

Sputter etching uses energetic noble gas ions such as Ar^+ to bombard the wafer surface. Etching occurs by physically knocking atoms off the surface of the wafer. Highly anisotropic etching can be obtained, but selectivity is often poor. Metals can be used as barrier materials to protect the wafer from etching.

Reactive-ion etching combines the plasma and sputter etching processes. Plasma systems are used to ionize reactive gases, and the ions are accelerated to bombard the surface. Etching occurs through a combination of the chemical reaction and momentum transfer from the etching species.

2.2.3 Photoresist Removal

After windows are etched through the SiO_2 layer, the photoresist is stripped from the surface, leaving a window in the silicon dioxide. Photoresist removal typically uses proprietary-liquid resist strippers, which cause the resist to swell and lose adhesion to the substrate. Dry processing may also be used to remove resist by oxidizing (burning) it in an oxygen plasma system, a process often called *resist ashing.*

2.3 PHOTOMASK FABRICATION

Photomask fabrication involves a series of photographic processes outlined in Fig. 2.6. An integrated-circuit mask begins with a large-scale drawing of each mask. Early photomasks were cut by hand in a material called *rubylith,* a sandwich of a clear backing layer and a thin red layer of Mylar. The red layer was cut with a stylus and peeled off, leaving the desired pattern in red. The original rubylith copy of the mask was 100 to 1000 times larger than the final integrated circuit and was photographically reduced to form a reticle for use in a step-and-repeat camera, as described later

Today, computer graphics systems and optical pattern generators have largely supplanted the use of rubylith. An image of the desired mask is created on a computer

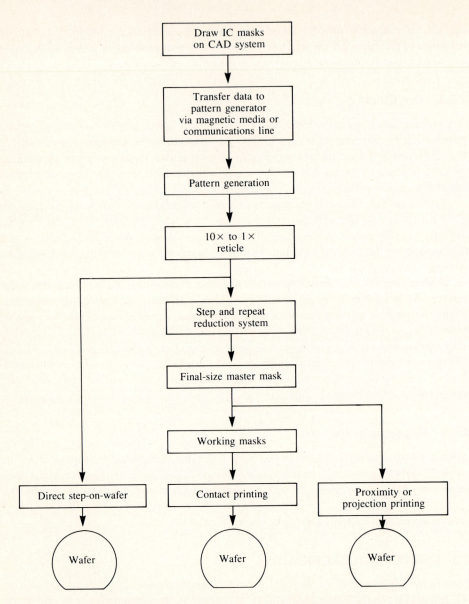

Fig. 2.6   Outline of steps in the mask fabrication process.

graphics system. Once the image is complete, files containing the commands needed to drive a pattern generator are created on magnetic tape or disks. The pattern generator uses a flash lamp to expose the series of rectangles composing the mask image directly onto a photographic plate called the *reticle*.

Reticle images range from one to ten times final size. A step-and-repeat camera is used to reduce the reticle image to its final size and to expose a two-dimensional array of images on a master copy of the final mask. On a 125-mm wafer, it is possible to get approximately 1900 copies of a 2.5 mm × 2.5 mm integrated-circuit chip! Figure 2.7 shows examples of a computer graphics plot, a reticle, and a final mask for a simple integrated circuit.

A final master copy of the mask is usually made in a thin film of metal, such as chrome, on a glass plate. The mask image is transferred to photoresist, which is used as an etch mask for the chrome. Working emulsion masks are then produced from the chrome master.

Each time a mask is brought into contact with the surface of the silicon wafer, the pattern can be damaged. Therefore, emulsion masks are used for only a few exposures before they are thrown away. Contact printing has been largely replaced by proximity and projection printing systems, illustrated in Fig. 2.8. In proximity printing, the mask is brought in very close proximity to the wafer but does not come in contact with the wafer during exposure, thus preventing damage to the mask. Projection printing uses a dual-lens system to project a portion of the mask image onto the wafer surface. The wafer and masks may be scanned or the system may operate in a step-and-repeat mode. The actual mask and lenses are mounted many centimeters from the wafer surface.

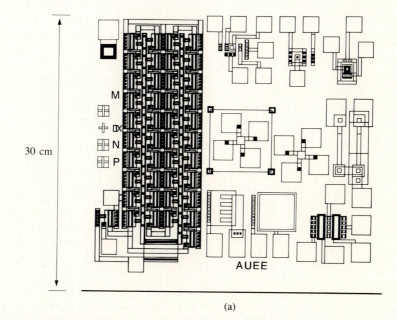

30 cm

(a)

Fig. 2.7 Mask fabrication. (a) Composite computer graphics plot of all masks for a simple integrated circuit; (b) 10× reticle of metal-level mask; (c) final-size emulsion mask with 400 copies of the metal level of the integrated circuit in (a). (Figure continued on p. 24.)

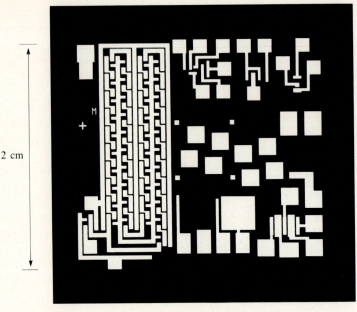

2 cm

(b)

10 cm

(c)

Fig. 2.7   (continued)

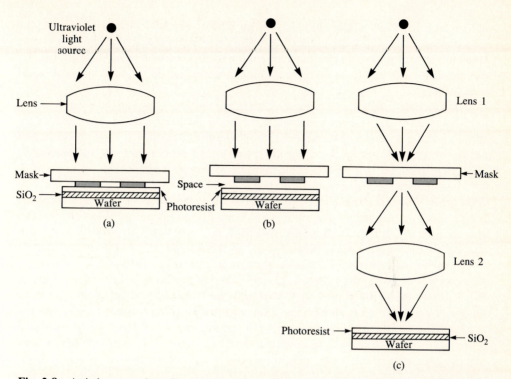

Fig. 2.8 Artist's conception of various printing techniques. (a) Contact printing, in which wafer is in intimate contact with mask; (b) proximity printing, in which wafer and mask are in close proximity; (c) projection printing, in which light source is scanned across the mask and focused on the wafer. Copyright, 1983, Bell Telephone Laboratories, Incorporated. Reprinted by permission from ref. [2].

In large-diameter wafers, it is difficult to maintain alignment between mask levels across the complete wafer, particularly with features whose size approaches 1 μm. High-resolution systems now use direct step-on-wafer techniques. A projection system is used with a 1× or 10× reticle to expose the integrated-circuit die pattern directly on the wafer. No step-and-repeat masks of the circuit are produced. The pattern is aligned and exposed separately at each die site.

Masks requiring geometrical features smaller than 1.25 μm can be produced by writing the pattern on the wafer in a special electron-sensitive resist using electron beams. Electron-beam systems are also commonly used to make 1× reticles for direct step-on-wafer systems.

2.4 SUMMARY

Photolithography is used to transfer patterns from masks to photoresist on the surface of silicon wafers. The resist protects portions of the surface while windows are etched in barrier layers such as silicon dioxide, silicon nitride, or metal. The windows may be etched using either wet or dry processing techniques. Wet chemical etching tends to etch under the edge of the mask, causing a loss of linewidth control at small dimensions. Dry etching can yield highly anisotropic etching profiles and is required in most VLSI processing.

After etching, impurities can be introduced into the wafer through the windows using ion implantation and/or high-temperature diffusion, or metal can be deposited on the surface making contact with the silicon through the etched windows. Masking operations are performed over and over during integrated-circuit processing, and the number of mask steps required is used as a basic measure of process complexity.

Mask fabrication uses computer graphics systems to draw the chip image at 100 to 2000 times final size. Reticles one to ten times final size are made from this computer image, using optical pattern generators or electron-beam systems. Step-and-repeat cameras are used to fabricate final masks from the reticles, or direct step-on-wafer systems may be used to transfer the patterns directly to the wafer.

Today we are reaching the limits of optical lithography. Present equipment can define windows which are approximately 1.25 μm wide. (Just a few year ago, experts thought that 2 μm would be the limit! Today it appears that it may be possible to extend optical lithography to submicron dimensions.) The wavelength of light is too long to produce much smaller geometrical features because of fringing and interference effects. Electron-beam and X-ray lithography are now being used to fabricate devices with geometrical features smaller than 0.25 μm, and lithography test structures have reproduced shapes with minimum feature sizes of 0.1 μm.

REFERENCES

[1] L. F. Thompson, C. G. Wilson, and M. J. Bowden, Eds., *Introduction to Microlithography,* American Chemical Society, Washington, D.C., 1983.

[2] S. M. Sze, Ed., *VLSI Technology,* McGraw-Hill, New York, 1983.

[3] D. J. Elliot, *Integrated Circuit Fabrication Technology,* McGraw-Hill, New York, 1982.

[4] W. Kern and D. A. Poutinen, "Cleaning Solutions Based upon Hydrogen Peroxide for Use in Silicon Semiconductor Technology," RCA Review, *31,* 187–206 (June, 1970).

[5] W. Kern, "Purifying Si and SiO$_2$ Surfaces with Hydrogen Peroxide," Semiconductor International, p. 94–99, (April, 1984).

FURTHER READING

1. M. C. King, "Principles of Optical Lithography," in *VLSI Electronics,* Vol. 1, N. G. Einspruch, Ed., Academic Press, New York, 1981.

2. J. H. Bruning, "A Tutorial on Optical Lithography," in *Semiconductor Technology,* D. A. Doane, D. B. Fraser, and D. W. Hess, Eds., Electrochemical Society, Pennington, NJ, 1982.

3. R. K. Watts and J. H. Bruning, "A Review of Fine-line Lithographic Techniques: Present and Future," Solid-State Technology, *24,* 99–105 (May, 1981).

4. J. A. Reynolds, "An Overview of E-Beam Mask-Making," Solid-State Technology, *22*, 87–94 (August 1979).

5. E. C. Douglas, "Advanced Process Technology for VLSI Circuits," Solid-State Technology, *24*, 65–72 (May, 1981).

6. L. M. Ephrath, "Etching Needs for VLSI," Solid-State Technology, *25*, 87–92 (July, 1982).

7. N. D. Wittels, "Fundamentals of Electron and X-ray Lithography," in *Fine Line Lithography*, R. Newman, Ed., North-Holland, Amsterdam, 1980.

8. D. Maydan, "X-ray Lithography for Microfabrication," Journal of Vacuum Science and Technology, *17*, 1164–1168 (September/October, 1980).

PROBLEMS

2.1 A complex CMOS fabrication process requires 15 masks. What fraction of the dice must be good (i.e., what yield must be obtained) during each mask step if we require 30% of the final dice to be good?

2.2 The mask set for a simple rectangular *pn* junction diode is shown in Fig. P2.2. The diode is formed in a *p*-type substrate. Draw a picture of the horizontal layout for the diode which results when a worst-case misalignment of 3 μm occurs in both the *x* and *y* directions on each mask level.

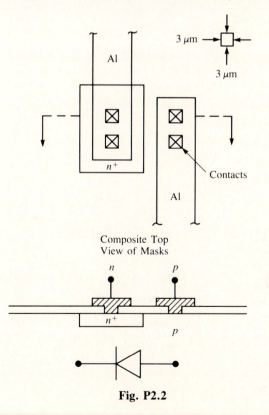

Fig. P2.2

(a) Assume that both the contact and metal levels are aligned to the diffusion level.

(b) Assume that the contact level is aligned to the diffusion level and the metal level is aligned to the contact level.

2.3 Figure P2.3 shows a resist pattern on top of a silicon dioxide film 1 μm thick. Draw the silicon–silicon dioxide structure after etching and removal of the photoresist for:

(a) Isotropic wet chemical etching

(b) Anisotropic dry etching with no undercutting.

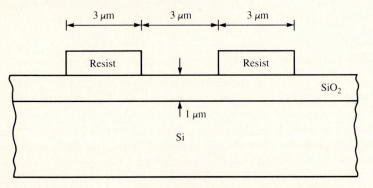

Fig. P2.3

2.4 (a) What type of photoresist must be used with each of the three mask levels (*n*-diffusion window, contact windows, and metal etch) used to fabricate the diode of Problem 2.2? Assume that the areas shown are dark on the mask (a "light-field mask").

(b) Draw a set of alignment marks suitable for use with the alignment sequence of Problem 2.2b.

3 / Thermal Oxidation of Silicon

Upon exposure to oxygen, the surface of a silicon wafer oxidizes to form silicon dioxide. This native silicon dioxide film is a high-quality electrical insulator and can be used as a barrier material during impurity diffusion. These two properties of silicon dioxide were the primary process factors leading to silicon becoming the dominant material in use today for the fabrication of integrated circuits. This chapter discusses the theory of oxide growth, oxide growth processes, factors affecting oxide growth rate, impurity redistribution during oxidation, and techniques for selective oxidation of silicon. Methods for determining the thickness of the oxide film are also presented.

3.1 THE OXIDATION PROCESS

Thermal oxidation of silicon is easily achieved by heating the wafer to a high temperature, typically 900 to 1200 °C, in an atmosphere containing either pure oxygen or water vapor. Both water vapor and oxygen move (diffuse) easily through silicon dioxide at these high temperatures (see Fig. 3.1). Oxygen arriving at the silicon surface can then combine with silicon to form silicon dioxide. The chemical reaction occurring at the silicon surface is

$$Si + O_2 \rightarrow SiO_2 \tag{3.1}$$

for dry oxygen and

$$Si + 2H_2O \rightarrow SiO_2 + 2H_2 \tag{3.2}$$

for water vapor. Silicon is consumed as the oxide grows, and the resulting oxide expands during growth, as shown in Fig. 3.2. The final oxide layer is approximately 54% above the original surface of the silicon and 46% below the original surface.

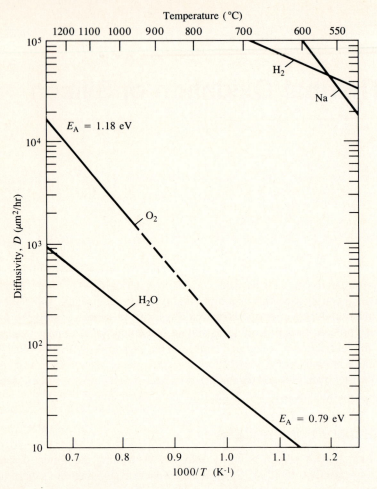

Fig. 3.1 Diffusivities of hydrogen, oxygen, sodium, and water vapor in silicon glass. Copyright John Wiley & Sons, Inc. Reprinted with permission from ref. [2].

3.2 MODELING OXIDATION

In order for oxidation to occur, oxygen must reach the silicon interface. As the oxide grows, oxygen must pass through more and more oxide, and the growth rate decreases as time goes on. A simple model for oxidation can be developed by assuming that oxygen diffuses through the existing oxide layer. *Fick's first law of diffusion* states that the particle flow per unit area, J (called *particle flux*), is directly proportional to the concentration gradient of the particle:

$$J = -D \, \partial N(x, t)/\partial x \tag{3.3}$$

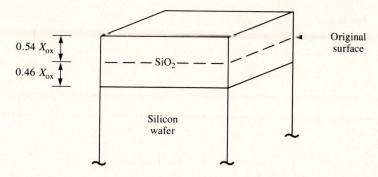

$0.54\ X_{ox}$

$0.46\ X_{ox}$

Original surface

SiO$_2$

Silicon wafer

Fig. 3.2 Formation of a silicon dioxide layer on the surface of a silicon wafer consumes silicon during growth of the layer. The oxide expands to fill a region approximately 54% above and 46% below the original surface of the wafer. The exact percentages depend on the density of the oxide.

where D is the diffusion coefficient and N is the particle concentration. The negative sign indicates that particles move from a region of high concentration to a region of low concentration.

For our case of silicon oxidation, we will make the approximation that the oxygen flux passing through the oxide in Fig. 3.3 is constant everywhere in the oxide (oxygen does not accumulate in the oxide). The oxygen flux J is then given by

$$J = -D(N_i - N_0)/X_0 \qquad \text{(number of particles/cm}^2\text{-sec)} \qquad (3.4)$$

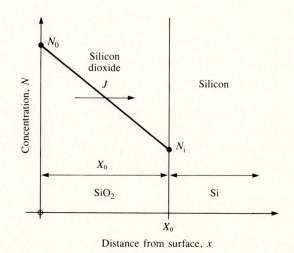

N_0

Silicon dioxide

J

Silicon

N_i

X_0

SiO$_2$ Si

Concentration, N

X_0

Distance from surface, x

Fig. 3.3 Model for thermal oxidation of silicon. X_0 is the thickness of the silicon dioxide layer at any time t. J is the constant flux of oxygen diffusing through the layer, and N_0 and N_i represent the oxygen concentration at the oxide surface and silicon dioxide–silicon interface, respectively.

where X_0 is the thickness of the oxide at a given time, and N_0 and N_i are the concentrations of the oxidizing species in the oxide at the oxide surface and silicon dioxide–silicon interface, respectively. At the silicon dioxide–silicon interface, we assume that the oxidation rate is proportional to the concentration of the oxidizing species so that the flux at the interface is

$$J = k_s N_i \tag{3.5}$$

where k_s is called the *rate constant* for the reaction at the Si-SiO$_2$ interface. Eliminating N_i using eqs. (3.4) and (3.5), the flux J becomes

$$J = DN_0/(X_0 + D/k_s) \tag{3.6}$$

The rate of change of thickness of the oxide layer with time is then given by the oxidizing flux divided by the number of molecules M of the oxidizing species that are incorporated into a unit volume of the resulting oxide:

$$dX_0/dt = J/M = (DN_0/M)/(X_0 + D/k_s) \tag{3.7}$$

This differential equation is easily solved using the boundary condition $X_0(t = 0) = X_i$, which yields

$$X_i^2 + AX_i = B\tau$$

or

$$t = X_0^2/B + X_0/(B/A) - \tau \tag{3.8}$$

where $A = 2D/k_s$, $B = 2DN_0/M$, and $\tau = X_i^2/B + X_i/(B/A)$. X_i is the initial thickness of oxide on the wafer. A thin native oxide layer (10 to 20 Å) is always present on silicon due to atmospheric oxidation, or X_i may represent a thicker oxide grown during previous oxidation steps. Solving eq. (3.8) for $X_0(t)$ yields

$$X_0(t) = 0.5A\left[\left\{1 + \frac{4B}{A^2}(t + \tau)\right\}^{1/2} - 1\right] \tag{3.9}$$

For short times with $(t + \tau) \ll A^2/4B$,

$$X_0(t) = (B/A)(t + \tau) \tag{3.10}$$

Oxide growth is proportional to time, and the ratio B/A is called the *linear (growth) rate constant*. In this region, growth rate is limited by the reaction at the silicon interface.

For long times with $(t + \tau) \gg A^2/4B, t \gg \tau$,

$$X_0 = \sqrt{Bt} \tag{3.11}$$

Oxide growth is proportional to the square root of time, and B is called the *parabolic rate constant*. The oxidation rate is diffusion-limited in this region.

3.3 FACTORS INFLUENCING OXIDATION RATE

There is good experimental agreement with this simple theory. Figures 3.4 and 3.5 show experimental data for the parabolic and linear rate constants. The rate-constant data follow straight lines when plotted on a semilogarithmic scale versus reciprocal temperature. This type of behavior occurs in many natural systems and is referred to as an *Arrhenius relationship*. A mathematical model for this behavior is as follows:

$$D = D_0 \exp(-E_A/kT) \tag{3.12}$$

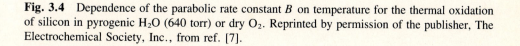

Fig. 3.4 Dependence of the parabolic rate constant B on temperature for the thermal oxidation of silicon in pyrogenic H_2O (640 torr) or dry O_2. Reprinted by permission of the publisher, The Electrochemical Society, Inc., from ref. [7].

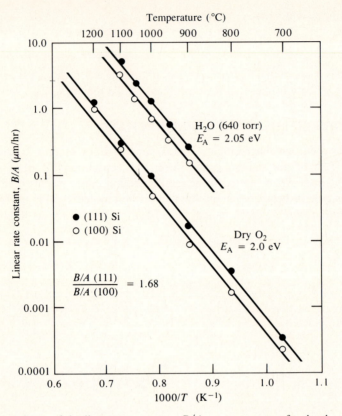

Fig. 3.5 Dependence of the linear rate constant B/A on temperature for the thermal oxidation of silicon in pyrogenic H_2O (640 torr) or dry O_2. Reprinted by permission of the publisher, The Electrochemical Society, Inc., from ref. [7].

Values for the coefficient D_0 and activation energy E_A for wet and dry oxygen are given in Table 3.1. For wet oxidation, a plot of the experimental data of oxide thickness versus oxidation time is consistent with an initial oxide thickness of approximately zero at $t = 0$. However, a similar plot for dry oxidation yields an initial oxide thickness of 250 Å for temperatures ranging from 800 to 1200 °C. Thus, a nonzero value for τ must be used in eq. (3.8) for dry oxidation calculations. This large value of X_i indicates that our simple oxidation theory is not quite correct, and the reason for this value of X_i is not well understood. Graphs of oxide growth versus time, calculated using the values from Table 3.1, are given in Figs. 3.6 and 3.7.

Eq. (3.12) indicates the strong dependence of oxide growth on temperature. A number of other factors affect the oxidation rate, including wet and dry oxidation, pressure, crystal orientation, and impurity doping. Water vapor has a much higher solubility than oxygen in silicon dioxide, which accounts for the much higher oxide

Table 3.1 Values for Coefficient D_0 and Activation Energy E_A for Wet and Dry Oxygen.*

	Wet O_2 ($X_i - 0$ nm)		Dry O_2 ($X_i = 25$ nm)	
	D_0	E_A	D_0	E_A
$\langle 100 \rangle$ Silicon				
Linear (B/A)....	9.70×10^7 μm/hr	2.05 eV	3.71×10^6 μm/hr	2.00 eV
Parabolic (B)....	386 μm^2/hr	0.78 eV	772 μm^2/hr	1.23 eV
$\langle 111 \rangle$ Silicon				
Linear (B/A)....	1.63×10^8 μm/hr	2.05 eV	6.23×10^6 μm/hr	2.00 eV
Parabolic (B)....	386 μm^2/hr	0.78 eV	772 μm^2/hr	1.23 eV

*Data from ref. [7].

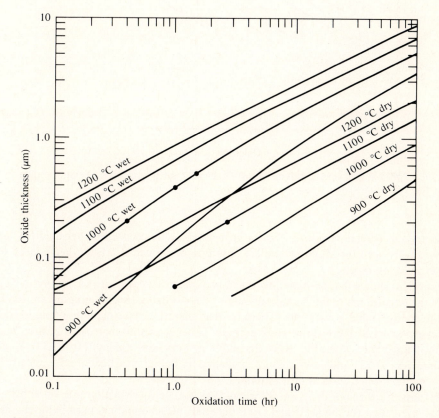

Fig. 3.6 Wet and dry silicon dioxide growth for $\langle 100 \rangle$ silicon calculated using the data from Table 3.1.

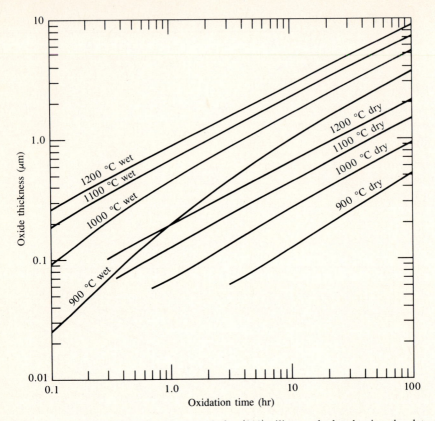

Fig. 3.7 Wet and dry silicon dioxide growth for ⟨111⟩ silicon calculated using the data from Table 3.1.

growth rate in a wet atmosphere. Slower growth results in a denser, higher-quality oxide and is usually used for MOS gate oxides. More-rapid growth in wet oxygen is used for thicker masking layers.

Eq. (3.8) shows that both the linear and parabolic rate constants are proportional to N_0. N_0 is proportional to the partial pressure of the oxidizing species, so pressure can be used to control oxide growth rate. There is great interest in developing low-temperature processes for VLSI fabrication. High pressure is being used to increase oxidation rates at low temperatures (see Fig. 3.8). In addition, very thin oxides (50 to 200 Å) are required for VLSI, and low-pressure oxidation is being investigated as a means of achieving controlled growth of thin oxides.

Figures 3.4 through 3.7 also show the dependence of oxidation rate on substrate crystal orientation for the ⟨111⟩ and ⟨100⟩ materials most commonly used in bipolar and MOS processes, respectively. The crystal orientation changes the number of silicon

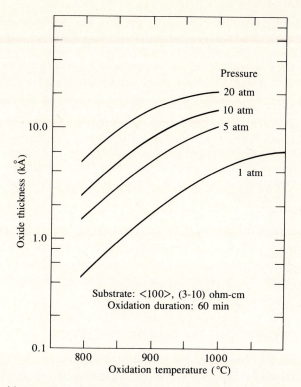

Fig. 3.8 Wet oxide growth at increased pressures. Reprinted with permission of Solid State Technology, published by Technical Publishing, a company of Dun and Bradstreet, from ref. [10].

bonds available at the silicon surface, which influences the oxide growth rate and quality of the silicon–silicon dioxide interface.

Example 3.1: Using Fig. 3.6, a 1-hr oxidation of ⟨100⟩ silicon at 1000 °C in dry oxygen will produce a silicon dioxide film approximately 580 Å (0.058 μm) thick. The same oxidation in wet oxygen will yield a film 3900 Å (0.39 μm) thick.

Example 3.2: A ⟨100⟩ wafer has a 2000-Å oxide on its surface. **(a)** How long did it take to grow this oxide at 1100 °C in dry oxygen? **(b)** The wafer is put back in the furnace in wet oxygen at 1000 °C. How long will it take to grow an additional 3000 Å of oxide? Solve this problem graphically using Fig. 3.6 and 3.7 as appropriate. **(c)** Repeat part (b) using the oxidation theory presented in eqs. (3.3) through (3.12).

Solution: (a) Using Fig. 3.6, it would take 2.75 hr to grow a 0.2-μm oxide in dry oxygen at 1100 °C.

(b) We can solve part (b) graphically using Fig. 3.6. The total oxide at the end of the oxidation would be 0.5 μm. If there were no oxide on the surface, it would take 1.5 hr to grow 0.5 μm. However, there is already a 0.2-μm oxide on the surface, and the furnace "thinks" that the wafer has already been in the furnace for 0.4 hr. The time required to grow the additional 0.3 μm of oxide is the difference in these two times: $\Delta t = (1.5 - 0.4)$ hr $= 1.1$ hr.

(c) From Table 3.1, $B = 3.86 \times 10^2 \exp(-0.78/kT)$ μm^2/hr and $(B/A) = 0.97 \times 10^8 \exp(-2.05/kT)$ μm/hr. Using $T = 1273$ K, $B = 0.314$ μm^2/hr and $(B/A) = 0.738$ μm/hr. Using these values and an initial oxide thickness of 0.2 μm yields a value of 0.398 hr for the effective initial oxidation time τ. Using τ and a final oxide thickness of 0.5 μm yields an oxidation time of 1.08 hr. Note that both the values of t and τ are close to those found in part (b). Of course, our results depend on our ability to interpolate logarithmic scales!

Heavy doping of silicon also changes its oxidation characteristics. Phosphorus doping increases the linear rate constant without altering the parabolic rate constant. Boron doping, on the other hand, increases the parabolic rate constant but has little effect on the linear rate constant. These effects are related to impurity redistribution during oxidation, which is discussed in the next section.

3.4 DOPANT REDISTRIBUTION DURING OXIDATION

During oxidation, the impurity concentration changes in the silicon near the silicon–silicon dioxide interface. Boron and gallium tend to be depleted from the surface, whereas phosphorus, arsenic, and antimony pile up at the surface.

Impurity depletion and pileup depend on both the diffusion coefficient and the *segregation coefficient* of the impurity in the oxide. The segregation coefficient m is equal to the ratio of the equilibrium concentration of the impurity in silicon to that of the impurity in the oxide. Various possibilities are depicted in Fig. 3.9. The value of m for boron is temperature-dependent and is less than 0.3 at normal diffusion temperatures. Boron also diffuses slowly through SiO$_2$. Thus, boron is depleted from the silicon surface and remains in the oxide (Fig. 3.9a). The presence of hydrogen during oxide growth or impurity diffusion greatly enhances the diffusion of boron through oxide, resulting in enhanced depletion of boron at the silicon surface (Fig. 3.9b).

The value of m is approximately ten for phosphorus, antimony, and arsenic. These elements are rejected by the oxide, and they diffuse slowly in the oxide, resulting in pileup at the silicon surface (Fig. 3.9c).

Gallium has a segregation coefficient of 20. However, it diffuses very rapidly through silicon dioxide. This combination causes depletion of gallium at the surface, as shown in Fig. 3.9d.

The effects of boron depletion and phosphorus pileup are particularly important in both bipolar and MOS processing. Process design must take both problems into account, and it may be necessary to add or change processing steps to overcome the effects of these phenomena.

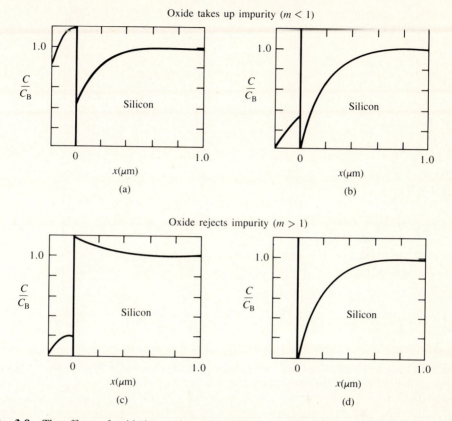

Fig. 3.9 The effects of oxidation on impurity profiles. (a) Slow diffusion in oxide (boron); (b) fast diffusion in oxide (boron with hydrogen ambient); (c) slow diffusion in oxide (phosphorus); (d) fast diffusion in oxide (gallium). Copyright John Wiley & Sons, Inc. Reprinted with permission from ref. [3].

3.5 MASKING PROPERTIES OF SILICON DIOXIDE

One of the most important properties of silicon dioxide is its ability to mask impurities during high-temperature diffusion. The diffusivities of antimony, arsenic, boron, and phosphorus in silicon dioxide are all orders of magnitude smaller than their corresponding values in silicon. Thus SiO_2 films can be used effectively to mask these elements. Relatively deep diffusion can take place in unprotected regions of silicon, whereas no significant impurity penetration will occur in regions covered by silicon dioxide.

Figure 3.10 gives the SiO_2 thickness required to mask boron and phosphorus diffusions as a function of diffusion time and temperature. Note that silicon dioxide is four to five times more effective in masking boron than in masking phosphorus. Arsenic and antimony diffuse more slowly than phosphorus, so an oxide thick enough to mask

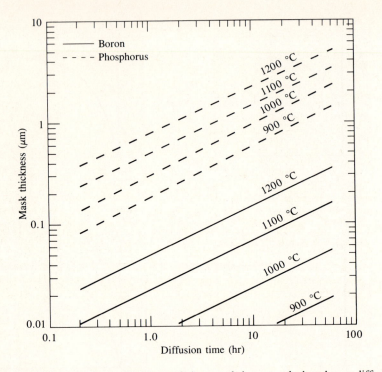

Fig. 3.10 Thickness of silicon dioxide needed to mask boron and phosphorus diffusions as a function of diffusion time and temperature.

phosphorus is also sufficient to mask arsenic and antimony. Masking oxide thicknesses of 0.5 to 1.0 μm are typical in integrated-circuit processes. The masking oxide would be considered to have failed if the impurity level under the mask were to reach a significant fraction (10%) of the background concentration in the silicon.

The graph for boron is valid for an environment which contains no hydrogen! As mentioned earlier, the presence of hydrogen greatly enhances the boron diffusivity. Wet oxidation releases hydrogen, and care must be taken to avoid boron diffusion in the presence of water vapor.

As mentioned in Section 3.4, gallium diffuses rapidly through SiO_2, as does aluminum, and silicon dioxide cannot be used as a mask for these elements. However, silicon nitride can be used effectively as a mask for these impurities.

3.6 TECHNOLOGY OF OXIDATION

Thermal oxidation of silicon is typically carried out in a high-temperature furnace tube. The tubes may be made of quartz, polycrystalline silicon, or silicon carbide and are

Fig. 3.11　A typical furnace used for oxidation and diffusion. This furnace contains six tubes, each with three-zone temperature control. Gases are supplied under automatic control from the rear of each tube.

specially fabricated to prevent sodium contamination during oxidation. The wafers are placed upright on edge in a quartz boat and pushed slowly into the furnace. The furnace is maintained at a temperature between 800 and 1200 °C. Three-zone resistance-heated furnaces maintain the temperature within a fraction of a degree over a distance of 0.5 m in the center zone. A photograph of a typical six-tube furnace used for oxidation and diffusion appears in Fig. 3.11.

The furnace is continually purged with an inert gas such as nitrogen prior to oxidation. Oxidation begins by introducing the oxidizing species into the furnace in gaseous form. Extremely high-purity oxygen is available and is used for dry oxidation. Water vapor may be introduced by passing oxygen through a bubbler containing deionized water heated to 95 °C. The oxygen serves as a transport gas to carry the water vapor into the furnace. High-purity water vapor can also be obtained by burning hydrogen and oxygen in the furnace tube. Steam is not often used because it tends to pit the silicon surface.

3.7　OXIDE QUALITY

Wet oxidation is used to grow relatively thick oxides used for masking. An oxidation growth cycle usually consists of a sequence of dry/wet/dry oxidations. Most of the oxide is grown during the wet oxidation phase since the growth rate is much higher in the presence of water. Dry oxidation results in a higher-density oxide than that achieved with wet oxidation. Higher density in turn results in a higher breakdown voltage (5 to 10 MV/cm). In order to maintain good process control, the thin gate oxides (<1000 Å) of MOS devices are usually formed using dry oxidation.

MOS devices are usually fabricated on wafers having a $\langle 100 \rangle$ surface orientation. The $\langle 100 \rangle$ orientation results in the smallest number of unsatisfied silicon bonds at the Si-SiO$_2$ interface, and the choice of the $\langle 100 \rangle$ orientation yields the lowest number of interface traps.*

Sodium ions are highly mobile in SiO$_2$ films (see Fig. 3.1), and contamination of MOS gate oxides was a difficult problem to overcome in the early days of the integrated-circuit industry.* Bipolar devices are much more tolerant of oxide contamination than MOS devices, and this was a primary factor in the early dominance of bipolar integrated circuits.

Sodium-ion contamination results in mobile positive charge in the oxide. In addition, a substantial level of positive fixed oxide exists at the Si-SiO$_2$ interface.* These charge centers attract electrons to the surface of MOS transistors, resulting in a negative shift in the threshold voltage of the MOS devices. NMOS devices become depletion-mode devices. PMOS devices remain enhancement-mode devices but have more negative threshold voltages. The first successful MOS processes were therefore PMOS processes. As the industry was able to improve overall oxide quality, NMOS processes became dominant because of the mobility advantage of electrons over holes.

It was discovered that the effects of sodium contamination can be greatly reduced by adding chlorine during oxidation. Chlorine is incorporated into the oxide and immobilizes the sodium ions. A small amount (6% or less) of anhydrous HCl can be added to the oxidizing gas. Gaseous chlorine, oxygen, or nitrogen can also be bubbled through trichloroethylene (C$_2$HCl$_3$). It should also be noted that the presence of chlorine during dry oxidation results in an increase in both the linear and parabolic rate constants.

3.8   SELECTIVE OXIDATION

The oxidation processes described above generally form an oxide film over the complete surface of the silicon wafer. The ability to selectively oxidize the silicon surface has become very important in high-density bipolar and MOS processes. Selective oxidation processes result in improved device packing density and more planar final structures.

Oxygen and water vapor do not diffuse well through silicon nitride. Figure 3.12 shows an MOS process using selective oxidation in which silicon nitride is used as an oxidation mask. A thin layer (10 to 20 nm) of silicon dioxide is first grown on the wafer to protect the silicon surface. Next, a layer of silicon nitride is deposited over the surface and patterned using photolithography. The wafer then goes through a thermal oxidation

---

*See Volume IV in the Modular Series on Solid State Devices, *Field Effect Devices,* Section 4.2, for an excellent discussion of oxide quality.

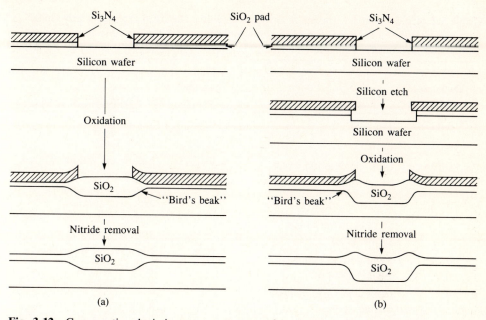

Fig. 3.12 Cross section depicting process sequence for (a) semirecessed and (b) fully recessed oxidations of silicon.

step. Oxide grows wherever the wafer is not protected by silicon nitride. This process results in the so-called *semirecessed oxide structure.*

Some oxide growth occurs under the edges of the nitride and causes the nitride to bend up at the edges of the masked area. The penetration of the oxide underneath the nitride results in a "bird's beak" structure. Formation of the bird's beak in Fig. 3.12 leads to loss of geometry control in VLSI structures, so minimization of the bird's beak phenomenon is an important goal in VLSI process design.

A *fully recessed oxide* can be formed by etching the silicon prior to oxidation. This process can yield a very planar surface after the removal of the nitride mask. Subsequent processing reduces the advantage of this process over the semirecessed version, and most processes today use some form of semirecessed oxidation.

3.9 OXIDE THICKNESS CHARACTERIZATION

One of the simplest methods for determining the thickness of an oxide is to compare the color of the wafer with the reference color chart in Table 3.2. When a wafer is illuminated with white light perpendicular to the surface, the light penetrates the oxide film and is reflected by the underlying silicon wafer. Constructive interference causes enhancement of a certain wavelength in the reflected light, and the color of the wafer corresponds to the enhanced wavelength. Constructive interference occurs when the

Table 3.2   Color Chart for Thermally Grown SiO_2 Films Observed Perpendicularly Under Daylight Fluorescent Lighting. Copyright 1964 by International Business Machines Corporation; reprinted with permission from ref. [9].

Film Thickness (μm)	Color and Comments	Film Thickness (μm)	Color and Comments
0.05	Tan	0.54	Yellow green
0.07	Brown	0.56	Green yellow
0.10	Dark violet to red violet	0.57	Yellow to "yellowish" (not yellow but is in the position where yellow is to be expected; at times appears to be light creamy gray or metallic)
0.12	Royal blue		
0.15	Light blue to metallic blue		
0.17	Metallic to very light yellow green		
0.20	Light gold or yellow; slightly metallic		
		0.58	Light orange or yellow to pink borderline
0.22	Gold with slight yellow orange	0.60	Carnation pink
		0.63	Violet red
0.25	Orange to melon	0.68	"Bluish" (not blue but borderline between violet and blue green; appears more like a mixture between violet red and blue green and looks grayish)
0.27	Red violet		
0.30	Blue to violet blue		
0.31	Blue		
0.32	Blue to blue green		
0.34	Light green	0.72	Blue green to green (quite broad)
0.35	Green to yellow green		
0.36	Yellow green	0.77	"Yellowish"
0.37	Green yellow	0.80	Orange (rather broad for orange)
0.39	Yellow		
0.41	Light orange	0.82	Salmon
0.42	Carnation pink	0.85	Dull, light red violet
0.44	Violet red	0.86	Violet
0.46	Red violet	0.87	Blue violet
0.47	Violet	0.89	Blue
0.48	Blue violet	0.92	Blue green
0.49	Blue	0.95	Dull yellow green
0.50	Blue green	0.97	Yellow to "yellowish"
0.52	Green (broad)	0.99	Orange
		1.00	Carnation pink

Table 3.2   (continued)

Film Thickness (μm)	Color and Comments	Film Thickness (μm)	Color and Comments
1.02	Violet red	1.24	Carnation pink to salmon
1.05	Red violet	1.25	Orange
1.06	Violet	1.28	"Yellowish"
1.07	Blue violet	1.32	Sky blue to green blue
1.10	Green	1.40	Orange
1.11	Yellow green	1.45	Violet
1.12	Green	1.46	Blue violet
1.18	Violet	1.50	Blue
1.19	Red violet	1.54	Dull yellow green
1.21	Violet red		

path length in the oxide ($2X_0$) is equal to an even multiple of one wavelength of light in the oxide.

$$2X_0 = k\lambda/n \qquad (3.13)$$

where the number k is any integer greater than zero, and n is the refractive index of the oxide ($n = 1.46$ for SiO_2).

As an example, a wafer with a 5000-Å silicon dioxide layer will appear blue green. Color-chart comparisons are quite subjective, and the colors vary periodically with thickness. In addition, care must be exercised to determine the color from a position perpendicular to the wafer. The color chart (Table 3.2) is only valid for vertical illumination with fluorescent light.

Accurate thickness measurement can be achieved with an instrument called an *ellipsometer,* and this instrument is often used to make an accurate reference color chart. Polarized monochromatic light is used to illuminate the wafer at an angle to the surface. Light is reflected from both the oxide and silicon surfaces. The differences in polarization are measured, and the oxide thickness can then be calculated.[15]

A mechanical surface profiler can also be used to measure film thickness. The oxide is partially etched from the surface of a test wafer to expose a step between the wafer and oxide surfaces. A stylus is mechanically scanned over the surface of the wafer, and thickness variations are recorded on a strip-chart recorder. Films ranging from less than 0.01 μm to more than 5 μm can be measured with this instrument.

Accurate film thickness measurements can also be achieved using light-interference effects in microscopy, and automated interference-based equipment is commercially available for thin-film characterization.

3.10   SUMMARY

Silicon dioxide provides a high-quality insulating barrier on the surface of the silicon wafer. In addition, this layer can serve as a barrier layer during subsequent impurity-diffusion process steps. These two factors have allowed silicon to become the dominant semiconductor material in use today.

A native oxide layer several tens of angstroms thick forms on the surface of silicon immediately upon exposure to oxygen even at room temperature. The thickness of this oxide layer may be readily measured from the accumulation-region capacitance of a MOS test capacitor. Thicker layers of silicon dioxide are conveniently grown in high-temperature oxidation furnaces using both wet and dry oxygen. Oxidation occurs much more rapidly in wet oxygen than in dry oxygen. However, the dry-oxygen environment produces a higher-quality oxide and is usually used for the growth of MOS gate oxides. Oxide cleanness is extremely important for MOS processes, and great care is exercised to prevent sodium contamination of the oxide. The addition of chlorine during oxidation improves oxide quality. Finally, oxidation alters the impurity distribution at the surface of the silicon wafer. Boron tends to be depleted from the silicon surface, whereas phosphorus tends to pile up at the silicon surface.

Oxidation thickness can be accurately measured using ellipsometers, interference microscopes, and mechanical surface profilers or can be estimated from the apparent color of the oxide under vertical illumination with white light.

REFERENCES

[1] S. M. Sze, *VLSI Technology,* McGraw-Hill, New York, 1983.

[2] S. K. Ghandhi, *VLSI Fabrication Principles,* John Wiley & Sons, New York, 1983.

[3] R. A. Colclaser, *Microelectronics—Processing and Device Design,* John Wiley & Sons, New York, 1980.

[4] W. R. Runyan, *Semiconductor Measurements and Instrumentation,* McGraw-Hill, New York, 1975.

[5] S. K. Ghandhi, *The Theory and Practice of Microelectronics,* John Wiley & Sons, New York, 1968.

[6] B. E. Deal and A. S. Grove, "General Relationship for the Thermal Oxidation of Silicon," Journal of Applied Physics, *36,* 3770–3778 (December, 1965).

[7] B. E. Deal, "Thermal Oxidation Kinetics of Silicon in Pyrogenic H_2O and 5% HCL/H_2O Mixtures," Journal of the Electrochemical Society, *125,* 576–579 (April, 1978).

[8] B. E. Deal, "The Oxidation of Silicon in Dry Oxygen, Wet Oxygen and Steam," Journal of the Electrochemical Society, *110,* 527–533 (June, 1963).

[9] W. A. Pliskin and E. E. Conrad, "Nondestructive Determination of Thickness and Refractive Index of Transparent Films," IBM Journal of Research & Development, *8,* 43–51 (January, 1964).

[10] S. C. Su, "Low Temperature Silicon Processing Techniques for VLSIC Fabrication," Solid-State Technology, *24*, 72–82 (March, 1981).

[11] A. S. Grove, O. Leistiko, and C. T. Sah, "Redistribution of Acceptor and Donor Impurities During Thermal Oxidation of Silicon," Journal of Applied Physics, *35*, 2695–2701 (September, 1964).

[12] A. S. Grove, *Physics and Technology of Semiconductor Devices,* John Wiley & Sons, New York, 1967.

[13] E. H. Nicollian and J. R. Brews, *MOS Physics and Technology,* John Wiley & Sons, New York, 1982.

[14] M. Ghezzo and D. M. Brown, "Diffusivity Summary of B, Ga, P, As and Sb in SiO_2," Journal of the Electrochemical Society, *120*, 146–148 (January, 1973).

[15] E. Passaglia, R. R. Stromberg, and J. Kruger, Eds., "Ellipsometry in the Measurement of Surfaces and Thin Films," National Bureau of Standards, Miscellaneous Publication #256, 1964.

PROBLEMS

3.1 How long does it take to grow 100 nm of oxide in wet oxygen at 1000 °C (assume $\langle 100 \rangle$ silicon)? In dry oxygen? Which process would be preferred?

3.2 A 1.2-μm silicon dioxide film is grown on a $\langle 100 \rangle$ silicon wafer in wet oxygen at 1100 °C. How long does it take to grow the first 0.4 μm? The second 0.4 μm? The final 0.4 μm?

3.3 Derive eq. (3.8) by solving differential eq. (3.7).

3.4 How much oxide is needed to mask a 4-hr boron diffusion at 1150 °C? A 1-hr phosphorus diffusion at 1050 °C?

3.5 A square window is etched through 200 nm of oxide prior to a second oxidation as in Example 3.2. The second oxidation grows 300 nm of oxide in the thick oxide region. Make a scale drawing of the cross section of the wafer after the second oxidation. What are the colors of the various regions under vertical illumination by white light?

3.6 Write a computer program to calculate the linear and parabolic rate constants for wet and dry oxidation for temperatures of 950, 1000, 1050, 1100, 1150, and 1200 °C. Assume $\langle 100 \rangle$ silicon.

3.7 A $\langle 100 \rangle$ silicon wafer has 400 nm of oxide on its surface. How long will it take to grow an additional 1 μm of oxide in wet oxygen at 1100 °C? Compare graphical and mathematical results. What is the color of the final oxide under vertical illumination by white light?

3.8 Yellow light has a wavelength of approximately 0.57 μm. Calculate the thicknesses of silicon dioxide which will appear yellow under vertical illumination by white light. Consider oxide thicknesses less than 1.5 μm. Compare with the color chart (Table 3.2).

3.9 Write a computer program to calculate the time required to grow a given thickness of oxide based on the theory of Section 3.2. The user should be able to specify desired oxide thickness, wet or dry oxidation conditions, temperature, and orientation of the silicon wafer.

4 / Diffusion

High-temperature diffusion has historically been one of the most important processing steps used in the fabrication of monolithic integrated circuits. Diffusion has been the primary method of introducing impurities such as boron, phosphorus, and antimony into silicon to control the majority-carrier type and resistivity of layers formed in the wafer. In this chapter, we explore the theoretical and practical aspects of the diffusion process, the characterization of diffused layer sheet resistance, and the determination of junction depth. Physical diffusion systems and solid, liquid, and gaseous impurity sources are all discussed.

4.1 THE DIFFUSION PROCESS

The diffusion process begins with the deposition of a high concentration of the desired impurity on the silicon surface through windows etched in the protective barrier layer. At high temperatures (900 to 1200 °C), the impurity atoms move from the surface into the silicon crystal via the *substitutional* or *interstitial* diffusion mechanisms illustrated in Fig. 4.1.

In the case of substitutional diffusion, the impurity atom hops from one crystal lattice site to another. The impurity atom thereby "substitutes" for a silicon atom in the lattice. Vacancies must be present in the silicon lattice in order for the substitutional process to occur. Statistically, a certain number of vacancies will always exist in the lattice. At high temperatures, vacancies may also be created by displacing silicon atoms from their normal lattice positions into the vacant *interstitial* space between lattice sites. The substitutional diffusion process in which silicon atoms are displaced into interstitial sites is called *interstitialcy* diffusion.

Considerable space exists between atoms in the silicon lattice, and certain impurity atoms diffuse through the crystal by jumping from one interstitial site to another. Since this mechanism does not require the presence of vacancies, interstitial diffusion proceeds much more rapidly than substitutional diffusion. The rapid diffusion rate makes interstitial diffusion difficult to control.

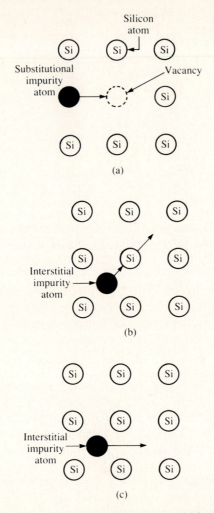

Fig. 4.1 Atomic diffusion in a two-dimensional lattice. (a) Substitutional diffusion, in which the impurity moves among vacancies in the lattice; (b) interstitialcy mechanism, in which the impurity atom replaces a silicon atom in the lattice, and the silicon atom is displaced to an interstitial site; (c) interstitial diffusion, in which impurity atoms do not replace atoms in the crystal lattice.

Impurity atoms need to occupy substitutional sites in the lattice in order to provide electrons or holes for conduction, as described in Volume I of this series.[1] Substitutional diffusion proceeds at a relatively low rate because the supply of vacancies is limited, but this slow diffusion rate is actually an advantage because it permits good control of the diffusion process.

4.2 MATHEMATICAL MODEL FOR DIFFUSION

The basic one-dimensional diffusion process follows *Fick's first law* of diffusion, presented in Chapter 3:

$$J = -D\, \partial N / \partial x \tag{4.1}$$

where J is the particle flux of the donor or acceptor impurity species, N is the concentration of the impurity, and D is the diffusion coefficient.

Fick's second law of diffusion may be derived using the continuity equation for the particle flux:

$$\partial N / \partial t = -\partial J / \partial x \tag{4.2}$$

Eq. (4.2) states that the rate of increase of concentration with time is equal to the negative of the divergence of the particle flux. For the one-dimensional case, the divergence is equal to the gradient. Combining eqs. (4.1) and (4.2) yields Fick's second law of diffusion:

$$\partial N / \partial t = D\, \partial^2 N / \partial x^2 \tag{4.3}$$

in which the diffusion coefficient D has been assumed to be independent of position. This assumption is violated at high impurity concentrations (see Section 4.8).

The partial differential equation in eq. (4.3) can be solved by variable separation or Laplace transform techniques. Two specific types of boundary conditions are important in modeling impurity diffusion in silicon. The first is the *constant-source diffusion*, in which the surface concentration is held constant throughout the diffusion. The second is called a *limited-source diffusion*, in which a fixed quantity of the impurity species is deposited in a thin layer on the surface of the silicon.

4.2.1 Constant-Source Diffusion

During a constant-source diffusion, the impurity concentration is held constant at the surface of the wafer. Under this boundary condition, the solution to eq. (4.3) is given by

$$N(x, t) = N_0\, \text{erfc}(x / 2\sqrt{Dt}) \tag{4.4}$$

for a semi-infinite wafer in which N_0 is the impurity concentration at the wafer surface ($x = 0$). Such a diffusion is called a *complementary error function* (erfc) diffusion, shown graphically in Fig. 4.2. As time progresses, the diffusion front proceeds further and further into the wafer with the surface concentration remaining constant. The total number of impurity atoms per unit area in the silicon is called the *dose, Q*, with units of atoms/cm^2. Q increases with time, and an external impurity source must supply a

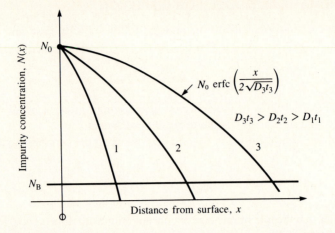

Fig. 4.2 A constant-source diffusion results in a complementary error function impurity distribution. The surface concentration N_0 remains constant and the diffusion moves deeper into the silicon wafer as the Dt product increases. Dt can change as a result of increasing diffusion time, increasing diffusion temperature, or a combination of both.

continual flow of impurity atoms to the surface of the wafer. The dose is found by integrating the diffused impurity concentration throughout the silicon wafer.

$$Q = \int_0^\infty N(x,t)\,dx = 2N_0\sqrt{Dt/\pi} \tag{4.5}$$

4.2.2 Limited-Source Diffusion

A limited-source diffusion is modeled mathematically using an impulse function at the silicon surface as the initial boundary condition. The magnitude of the impulse is equal to the dose Q. For this boundary condition in a semi-infinite wafer, the solution to eq. (4.3) is given by the Gaussian distribution,

$$N(x,t) = (Q/\sqrt{\pi Dt})\exp{-(x/2\sqrt{Dt})^2} \tag{4.6}$$

which is displayed graphically in Fig. 4.3. The dose remains constant throughout the limited-source diffusion process. As the diffusion front moves into the wafer, the surface concentration must decrease so that the area under the curve can remain constant with time.

On a normalized logarithmic plot, the shapes of the Gaussian and complementary error function curves appear similar, as illustrated in Fig. 4.4. The erfc curve, however, falls off more rapidly than the Gaussian curve.

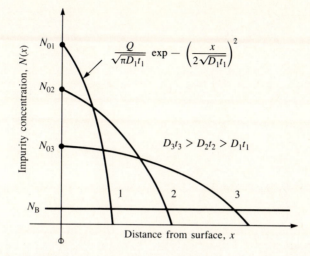

Fig. 4.3 A Gaussian distribution results from a limited-source diffusion. As the Dt product increases, the diffusion front moves more deeply into the wafer and the surface concentration decreases. The area under each of the three curves is the same.

4.2.3 Two-Step Diffusion

A short constant-source diffusion is often followed by a limited-source diffusion, resulting in a "two-step" diffusion process. The constant-source diffusion step is used to establish a known dose in a shallow layer on the surface of the silicon and is called the *predeposition* step. The fixed dose approximates an impulse and serves as the impurity source for the second diffusion step.

The second diffusion is called the *drive-in* step and is used to move the diffusion front to the desired depth. If the Dt product for the drive-in step is much greater than the Dt product for the predeposition step, the resulting impurity profile is closely approximated by a Gaussian distribution, eq. (4.6). If the Dt product for the drive-in step is much less than the Dt product for the predeposition step, the resulting impurity profile is closely approximated by a complementary error function distribution, eq. (4.4). An integral equation solution to the diffusion equation also exists for diffusion conditions which do not satisfy either inequality. (See ref. [25].)

4.3 THE DIFFUSION COEFFICIENT

Figure 4.5 shows the temperature dependence of the diffusion coefficient D for substitutional and interstitial diffusers in silicon. The large difference between these coefficients is readily apparent. In order to achieve reasonable diffusion times with substitutional diffusers, temperatures in the range of 900 to 1200 °C are typically used. Interstitial

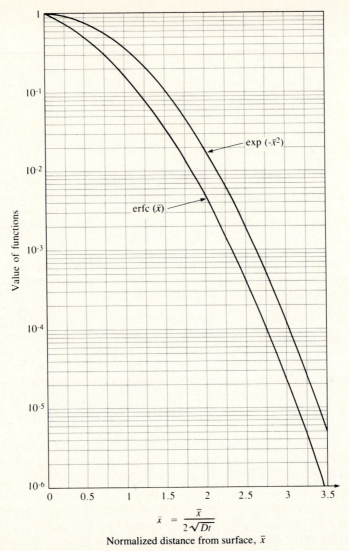

$$\bar{x} = \frac{x}{2\sqrt{Dt}}$$

Normalized distance from surface, $\bar{x}$

Fig. 4.4 A graph comparing the Gaussian and complementary error function (erfc) profiles. We will use this curve to evaluate the erfc and its inverse.

diffusers are difficult to control because of their large diffusion coefficients (see Problem 4.13).

Diffusion coefficients depend exponentially on temperature and follow the Arrhenius behavior discussed in Chapter 3:

$$D = D_0 \exp(-E_A/kT) \tag{4.7}$$

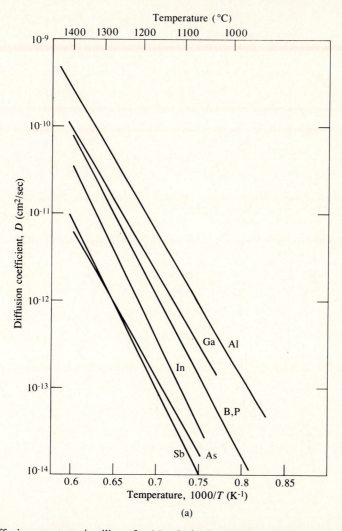

(a)

Fig. 4.5 Diffusion constants in silicon for (a) substitutional diffusers (above) and (b) interstitial diffusers (next page). Copyright John Wiley & Sons, Inc; reprinted with permission from ref. [25].

Values for D_0 and E_A can be determined from Fig. 4.5. Typical values for a number of impurities are given in Table 4.1.

Wide variability exists in diffusion coefficient data reported in the literature. We will use eq. (4.7) and Table 4.1 in the examples and problems throughout the rest of this book. In general, calculations based on eq. (4.7) and Table 4.1 can be used as guides. Most processes are then experimentally calibrated under the specific diffusion conditions in each laboratory.

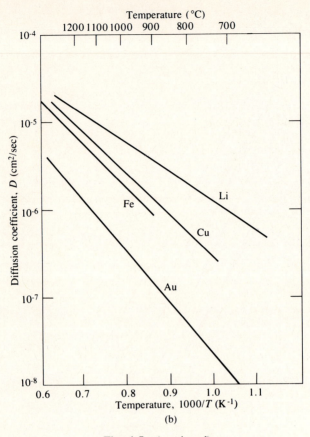

Fig. 4.5   (continued)

Table 4.1   Typical Diffusion Coefficient Values for a Number of Impurities.

Element	$D_0(cm^2/sec)$	$E_A(eV)$
B	10.5	3.69
Al	8.00	3.47
Ga	3.60	3.51
In	16.5	3.90
P	10.5	3.69
As	0.32	3.56
Sb	5.60	3.95

Example 4.1: Calculate the diffusion coefficient for boron at 1100 °C.

Solution: From Table 4.1, $D_0 = 10.5$ cm^2/sec and $E_A = 3.69$ eV. $T = 1373$ K. $D = 10.5$ exp $-(3.69/(8.614 \times 10^{-5})(1373)) = 2.96 \times 10^{-13}$ cm^2/sec.

4.4 SUCCESSIVE DIFFUSIONS

We are ultimately interested in the final impurity distribution after all processing is complete. A wafer typically goes through many time-temperature cycles during predeposition, drive-in, oxide growth, CVD, etc. For example, the base diffusion in a bipolar transistor will be followed by several high-temperature oxidations as well as the emitter predeposition and drive-in cycles. These steps take place at different temperatures for different lengths of time. The effect of these steps is determined by calculating the total Dt product, $(Dt)_{tot}$, for the diffusion. $(Dt)_{tot}$ is equal to the sum of the Dt products for all high-temperature cycles affecting the diffusion:

$$(Dt)_{tot} = \sum_i D_i t_i \qquad (4.8)$$

D_i and t_i are the diffusion coefficient and time associated with the ith processing step. $(Dt)_{tot}$ is then used in eq. (4.4) or (4.6) to determine the final impurity distribution.

4.5 SOLID-SOLUBILITY LIMITS

At a given temperature, there is an upper limit to the amount of an impurity which can be absorbed by silicon. This quantity is called the *solid-solubility limit* for the impurity and is indicated by the solid lines in Fig. 4.6 for boron, phosphorus, antimony, and arsenic at normal diffusion temperatures. As can be seen in Fig. 4.6, surface concentrations achieved through solid-solubility-limited diffusions will be quite high. For example, the solid-solubility limit of boron is approximately 3.3×10^{20}/cm^3 at 1100 °C, and 1.2×10^{21}/cm^3 for phosphorus at the same temperature. High concentrations are desired for the emitter and subcollector diffusions in bipolar transistors and the source and drain diffusions in MOSFETs. However, solid-solubility-limited concentrations are too heavy for the base regions of bipolar transistors and for many resistors. The two-step diffusion process described in Section 4.2.3 was developed to overcome this problem.

At high concentrations, only a fraction of the impurities actually contribute holes or electrons for conduction. The dotted lines in Fig. 4.6 show the "electrically active" portion of the impurity concentration. These curves will be referred to again in Section 4.7.2.

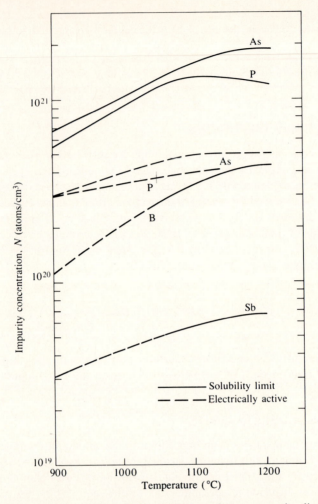

Fig. 4.6   The solid-solubility and electrically active impurity-concentration limits in silicon for antimony, arsenic, boron, and phosphorus. Reprinted with permission from ref. [26]. This paper was originally presented at the 1977 Spring Meeting of The Electrochemical Society, Inc., held in Philadelphia, Pennsylvania.

4.6   JUNCTION FORMATION AND CHARACTERIZATION

4.6.1   Vertical Diffusion and Junction Formation

The goal of most diffusions is to form *pn* junctions by converting *p*-type material to *n*-type material or vice versa. In Fig. 4.7, for example, the wafer is uniformly doped *n*-type material with a concentration indicated by N_B, and the diffusing impurity is boron.

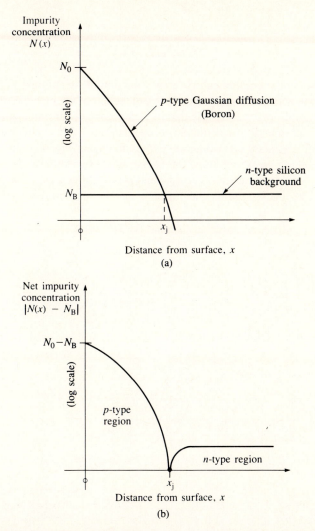

Fig. 4.7   Formation of a *pn* junction by diffusion. (a) An example of a *p*-type Gaussian diffusion into a uniformly doped *n*-type wafer; (b) net impurity concentration in the wafer. The metallurgical junction occurs at the point $x = x_j$, where the net concentration is zero. The material is converted to *p*-type to the left of x_j and remains *n*-type to the right of x_j.

The point at which the diffused impurity profile intersects the background concentration is the *metallurgical junction depth*, x_j. The net impurity concentration at x_j is zero. Setting $N(x)$ equal to the background concentration N_B at $x = x_j$ yields

$$x_j = 2\sqrt{Dt \; \ln(N_0/N_B)} \qquad (4.9a)$$

and

$$x_j = 2\sqrt{Dt}\ \text{erfc}^{-1}(N_B/N_0) \tag{4.9b}$$

for the Gaussian and complementary error function distributions, respectively. In Fig. 4.7, the boron concentration N exceeds N_B to the left of the junction, and this region is p-type. To the right of x_j, N is less than N_B, and this region remains n-type.

We can use our scientific calculators to evaluate eq. (4.9a), and we will learn to evaluate the complementary error function expression using Fig. 4.4. In order to calculate the junction depth, we must know the background concentration N_B of the original wafer. Figure 4.8 gives the resistivity of n- and p-type silicon as a function of doping concentration. The background concentration can be determined using this figure

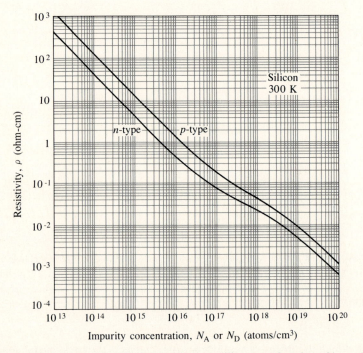

Fig. 4.8 Room-temperature resistivity in n- and p-type silicon as a function of impurity concentration. (Note that these curves are valid for either donor or acceptor impurities but not for compensated material containing both types of impurities.) Copyright 1987 Addison-Wesley Publishing Company. Reprinted with permission from ref. [3].

when uniform concentrations of either donor or acceptor impurities are present in the silicon wafer.

Example 4.2: A boron diffusion is used to form the base region of an *npn* transistor in a 0.18-ohm-cm *n*-type silicon wafer. A solid-solubility-limited boron predeposition is performed at 900 °C for 15 min followed by a 5-hr drive-in at 1100 °C. Find the surface concentration and junction depth (**a**) following the predeposition step and (**b**) following the drive-in step.

Solution: The predeposition step is a solid-solubility-limited constant-source diffusion. Using Fig. 4.6, the boron surface concentration is approximately $1.1 \times 10^{20}/\text{cm}^3$. The temperature of 900 °C equals 1173 K, which yields a diffusion coefficient $D_1 = 1.45 \times 10^{-15} \text{ cm}^2/\text{sec}$, and $t_1 = 900$ sec. The constant-source diffusion results in an erfc profile, and the impurity profile following predeposition is given by

$$N(x) = 1.1 \times 10^{20} \text{ erfc}(x/2\sqrt{D_1 t_1}) \text{ boron atoms/cm}^3$$

To find the junction depth x_j, we must find the point at which the concentration $N(x)$ is equal to the background concentration N_B. Using Fig. 4.8, we find that a 0.18-ohm-cm *n*-type wafer corresponds to a doping concentration of $3 \times 10^{16}/\text{cm}^3$. Thus,

$$1.1 \times 10^{20} \text{ erfc}(x_j/2\sqrt{D_1 t_1}) = 3 \times 10^{16}$$

Solving for x_j yields

$$x_j = 2\sqrt{D_1 t_1} \text{ erfc}^{-1}(0.000273) = 2(\sqrt{1.31 \times 10^{-12}})(2.57) \text{ cm} = 0.0587 \ \mu\text{m}$$

The dose in silicon is needed for the drive-in step and is equal to

$$Q = 2N_0\sqrt{D_1 t_1/\pi} = 2(1.1 \times 10^{20})\sqrt{(1.45 \times 10^{-15})(900)/\pi} \text{ boron atoms/cm}^2$$
$$Q = 1.42 \times 10^{14} \text{ boron atoms/cm}^2$$

At the drive-in temperature of 1100 °C (1373 K), $D_2 = 2.96 \times 10^{-13} \text{ cm}^2/\text{sec}$, and the drive-in time of 5 hr = 18,000 sec. Assuming that a Gaussian profile results from the drive-in step, the final profile is given by

$$N(x) = 1.1 \times 10^{18} \exp -(x/2\sqrt{D_2 t_2})^2 \text{ boron atoms/cm}^3 \qquad (4.2.1)$$

Setting eq. (4.2.1) equal to the background concentration yields the final junction depth of 2.77 μm. Figure 4.9 shows the concentrations at various points in the diffusion process.

We must check our assumption that the drive-in step results in a Gaussian profile. The Dt product for the predeposition step is $1.31 \times 10^{-12} \text{ cm}^2$, and the Dt product for the drive-in step is $5.33 \times 10^{-9} \text{ cm}^2$. Thus $D_2 t_2 \gg D_1 t_1$, and our assumption is justified.

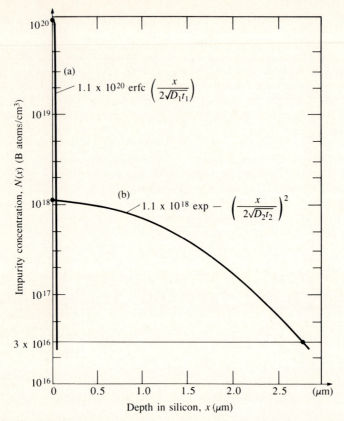

Fig. 4.9 Calculated boron impurity profiles for Example 4.2. (a) Following the predeposition step at 900 °C for 15 min; (b) following a subsequent 5-hr drive-in step at 1100 °C. The final junction depth is 2.77 μm with a surface concentration of 1.1×10^{18}/cm^3. The initial profile approximates an impulse.

4.6.2 Lateral Diffusion

During diffusion, impurities not only diffuse vertically but also move laterally under the edge of any diffusion barrier. Figure 4.10 presents the results of computer simulation of the two-dimensional diffusion process. The normalized impurity concentrations can be used to find the ratio of lateral to vertical diffusion. Lateral diffusion is an important effect-coupling device and process design and was an important factor driving the development of self-aligned polysilicon-gate MOS processes. The interaction of lateral diffusion and device layout will be discussed in greater detail in Chapters 9 and 10.

Example 4.3: An erfc diffusion results in a junction depth of 2 μm and a surface concentration of 1×10^{20}/cm^3. The background concentration of the wafer is 1×10^{16}/cm^3. What is the lateral diffusion underneath the edge of the mask?

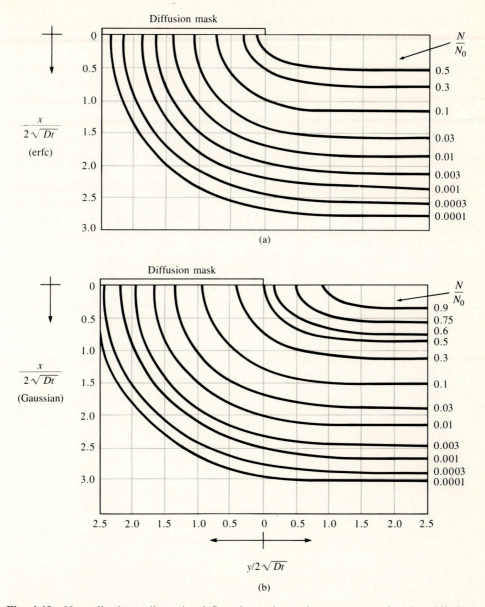

Fig. 4.10 Normalized two-dimensional Gaussian and complementary error function diffusions near the edge of a window in the barrier layer. Copyright 1965 by International Business Machines Corporation; reprinted with permission from ref. [4].

Solution: The junction occurs at $N = N_B$, and $N_0/N_B = 10^4$.

Using Fig. 4.10a, the ratio of lateral diffusion to vertical diffusion is 2.4/2.75, or 0.87. The lateral junction depth is therefore 1.74 μm.

4.6.3 Junction-Depth Measurement

Test wafers are normally processed in parallel with the actual integrated-circuit wafers. No masking is done on the test wafer so that diffusion may take place across its full surface. The test wafer provides a large area for experimental characterization of junction depth.

Two methods are commonly used to measure the junction depth of diffused layers. In the first, known as the *groove-and-stain* method, a cylindrical groove is mechanically ground into the surface of the wafer, as in Fig. 4.11. If the radius R of the grinding tool is known, the junction depth x is easily found to be

$$x_j = \sqrt{(R^2 - b^2)} - \sqrt{(R^2 - a^2)} \tag{4.10}$$

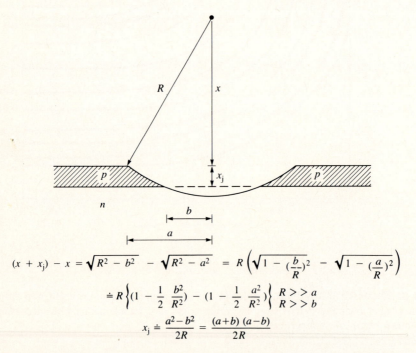

$$(x + x_j) - x = \sqrt{R^2 - b^2} - \sqrt{R^2 - a^2} = R\left(\sqrt{1 - (\tfrac{b}{R})^2} - \sqrt{1 - (\tfrac{a}{R})^2}\right)$$

$$\doteq R\left\{(1 - \tfrac{1}{2}\tfrac{b^2}{R^2}) - (1 - \tfrac{1}{2}\tfrac{a^2}{R^2})\right\} \quad \begin{matrix} R \gg a \\ R \gg b \end{matrix}$$

$$x_j \doteq \frac{a^2 - b^2}{2R} = \frac{(a+b)\,(a-b)}{2R}$$

Fig. 4.11 Junction-depth measurement by the groove-and-stain technique. The distances a and b are measured through a microscope, and the junction depth is calculated using eq. (4.11).

If the radius R is much larger than both distances a and b, then the junction depth is given approximately by

$$x_j = (a^2 - b^2)/2R = (a + b)(a - b)/2R \qquad (4.11)$$

After the grooving operation, the junction is delineated using a chemical etchant which stains the *pn* junction. Concentrated hydrofluoric acid with 0.1 to 0.5% nitric acid can be used as a stain which is enhanced through exposure to high-intensity light.[12] The distances a and b are measured through a microscope, and the junction depth is calculated using eq. (4.11).

The second technique is the *angle-lap* method. A piece of the wafer is mounted on a special fixture which permits the edge of the wafer to be lapped at an angle between 1 and 5°, as depicted in Fig. 4.12. The junction depth is magnified so that the distance on the lapped surface is given by

$$x_j = d \tan \theta = N\lambda/2 \qquad (4.12)$$

where θ is the angle of the fixture. An optically flat piece of glass is placed over the lapped region, and the test structure is illuminated with a collimated monochromatic

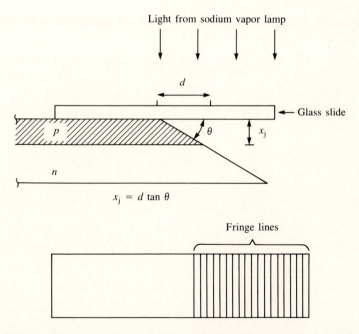

Fig. 4.12 Junction depth measurement by the angle-lap and stain method. Interference fringe lines are used to measure the distance d, which is related to the junction depth using eq. (4.12).

beam of light with wavelength λ, typically from a sodium vapor lamp. The resulting interference pattern has fringe lines which are approximately 0.29 μm apart. The number of fringes is counted through a microscope, and the junction depth may be found using eq. (4.12).

4.7   SHEET RESISTANCE

In diffused layers, resistivity is a strong function of depth. For circuit and device design, it is convenient to work with a new parameter, R_s, called *sheet resistance*, which eliminates the need to know the details of the diffused-layer profile.

4.7.1   Sheet-Resistance Definition

Let us first consider the resistance R of the rectangular block of uniformly doped material in Fig. 4.13. R is given by

$$R = \rho L/A \tag{4.13}$$

where ρ is the material's resistivity, and L and A represent the length and cross-sectional area of the block, respectively. Resistance is proportional to the material resistivity. If the length of the block is made longer, the resistance increases, and the resistance is inversely proportional to cross-sectional area.

Using W as the width of the sample and t as the thickness of the sample, the resistance may be rewritten as

$$R = (\rho/t)(L/W) = R_s(L/W) \tag{4.14}$$

where $R_s = (\rho/t)$ is called the *sheet resistance* of the layer of material. Given the sheet resistance R_s, a circuit designer need specify only the length and width of the resistor to

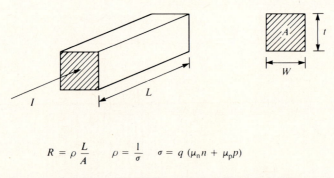

$$R = \rho\,\frac{L}{A} \qquad \rho = \frac{1}{\sigma} \qquad \sigma = q\,(\mu_n n + \mu_p p)$$

Fig. 4.13   Resistance of a block of material having uniform resistivity. A uniform current distribution is entering the material perpendicular to the end of the block. The ratio of resistivity to thickness is called the *sheet resistance* of the material.

define its value. Strictly speaking, the unit for sheet resistance is the ohm, since the ratio L/W is unitless. To avoid confusion between R and R_s, sheet resistance is given the special descriptive unit of ohms per square. The ratio L/W can be interpreted as the number of unit squares of material in the resistor.

Figure 4.14 shows top and side views of two typical dumbbell-shaped resistors with top contacts at the ends. The body of each resistor is seven "squares" long. If the sheet resistance of the diffusion were 50 ohms per square, each resistor would have a resistance of 350 ohms. Each end of the resistor adds approximately 0.65 squares to the resistor, and the total resistance would be approximately 415 ohms. Figure 4.15 gives the number of squares contributed by various end and corner configurations.

4.7.2   Irvin's Curves

From Section 4.2 we know that the impurity concentration resulting from a diffusion varies rapidly between the surface and the junction. Thus ρ is a function of depth for diffused resistors. For diffused layers, we define the sheet resistance R_s by

$$R_s = \overline{\rho}/x_j = \left[\int_0^{x_j} \sigma(x)\, dx \right]^{-1}$$

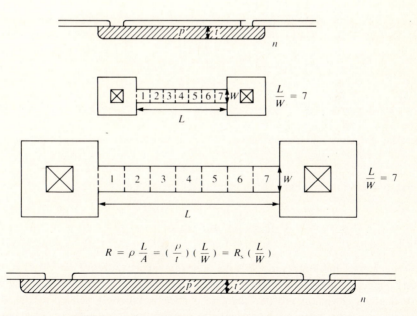

$$R = \rho \frac{L}{A} = \left(\frac{\rho}{t} \right) \left(\frac{L}{W} \right) = R_s \left(\frac{L}{W} \right)$$

Fig. 4.14   Top and side views of two diffused resistors of different physical size having equal values of resistance. Each resistor has a ratio L/W equal to 7 squares. Each end of the resistor contributes approximately 0.65 additional squares.

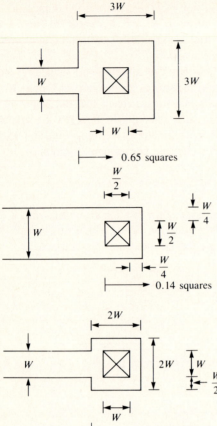

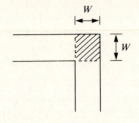

Corner = 0.56 squares

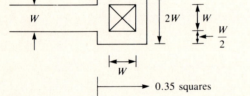

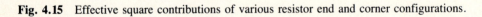

Fig. 4.15 Effective square contributions of various resistor end and corner configurations.

In extrinsic material, this expression can be approximated by

$$R_s = \left[\int_0^{x_j} q\mu N(x)\,dx \right]^{-1} \tag{4.15}$$

in which x_j is the junction depth, μ is the majority-carrier mobility, and $N(x)$ is the net impurity concentration. We neglect the depletion of charge carriers near the junction x_j.

For a given diffusion profile, sheet resistance is uniquely related to the surface concentration of the diffused layer and the background concentration of the wafer. Eq. (4.15) was evaluated numerically by Irvin,[5] and a number of Irvin's results have been combined into Figs. 4.16a–d.[2] These figures plot surface concentration versus the

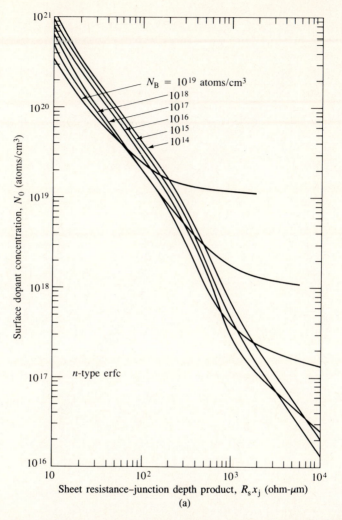

Fig. 4.16 Surface impurity concentration versus the sheet resistance–junction depth product for different silicon background concentrations at 300 K. (a) n-type erfc distribution; (b) n-type Gaussian distribution; (c) p-type erfc distribution; (d) p-type Gaussian distribution. After ref. [2]. Reprinted from ref. [5] with permission from the AT&T Technical Journal. Copyright 1962 AT&T. (This figure continues on pages 70, 71, and 72.)

$R_s x_j$ product and are used to find the sheet resistance and surface concentration of diffused layers.

Example 4.4: Find the sheet resistance of the base diffusion from Example 4.2.

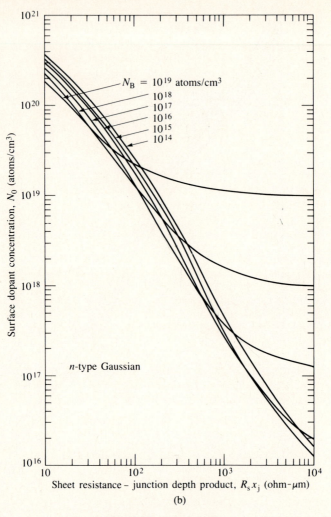

Fig. 4.16   (continued)

Solution: From Example 4.2, the background concentration of the wafer is $3 \times 10^{16}/\text{cm}^3$, the surface concentration is $1.1 \times 10^{18}/\text{cm}^3$, and the junction depth is 2.77 μm. The diffusion resulted in a p-type Gaussian layer. Using Fig. 4.16d, the $R_s x_j$ product is found to be approximately 800 ohm-μm. Dividing by a junction depth of 2.77 μm yields a sheet resistance of 289 ohms/square.

Sheet resistance is an electrical quantity which depends on the majority-carrier concentration. As shown in Fig. 4.6, the electrically active impurity concentration for

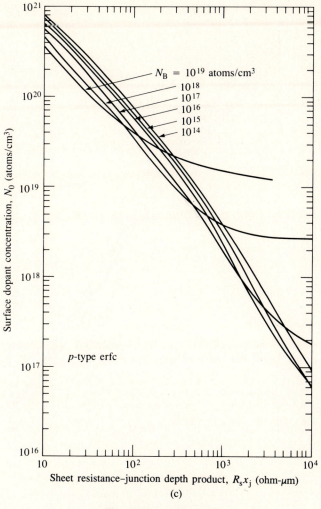

Fig. 4.16   (continued)

phosphorus and arsenic is considerably less than the total impurity concentration at high doping levels. In order to use Irvin's curves at high doping levels, the vertical axis, which is labeled "surface dopant density," should be interpreted to be the electrically active dopant concentration at the surface.

4.7.3   The Four-Point Probe

A special instrument called a *four-point probe* may be used to measure the bulk resistivity of starting wafers and the sheet resistance of shallow diffused layers. As shown

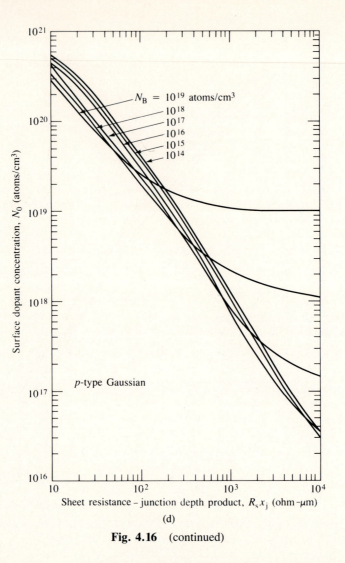

Fig. 4.16   (continued)

schematically in Fig. 4.17, a fixed current is injected into the wafer through the two outer probes, and the resulting voltage is measured between the two inner probes. If probes with a uniform spacing s are placed on an infinite slab of material, then the resistivity is given by

$$\rho = 2\pi s V/I \text{ ohm-meters for } t \gg s \tag{4.16}$$

and

$$\rho = (\pi t/\ln 2)V/I \text{ ohm-meters for } s \gg t \tag{4.17}$$

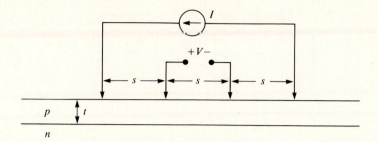

Fig. 4.17 Four-point probe with probe spacing s used for direct measurement of bulk wafer resistivity and the sheet resistance of thin diffused layers. A known current is forced through the outer probes, and the voltage developed is measured across the inner probes. (See eqs. (4.16) through (4.18).)

For shallow layers, eq. (4.17) gives the sheet resistance as

$$R_s = \rho/t = (\pi/\ln 2)V/I = 4.53 \, V/I \text{ ohm-meters for } s \gg t \qquad (4.18)$$

The approximation used in eqs. (4.17) and (4.18) is easily met for shallow diffused layers in silicon. Unfortunately, silicon wafers are often thinner than the probe spacing s, and the approximation in eq. (4.16) is not valid. Correction factors are given in Fig. 4.18 for thin wafers and for small-diameter wafers.[12]

4.7.4 Van der Pauw's Method

The sheet resistance of an arbitrarily shaped sample of material may be measured by placing four contacts on the periphery of the sample. A current is injected through one pair of the contacts, and the voltage is measured across another pair of contacts. Van der Pauw[13, 14] demonstrated that two of these measurements can be related by eq. (4.19) below.

$$\exp(-\pi t R_{AB,CD}/\rho) + \exp(-\pi t R_{BC,DA}/\rho) = 1 \qquad (4.19)$$

where $R_{AB,CD} = V_{CD}/I_{AB}$ and $R_{BC,DA} = V_{DA}/I_{BC}$. For a symmetrical structure like a square or a circle,

$$R_{AB,CD} = R_{BC,DA}$$

and

$$R_s = \rho/t = (\pi/\ln 2)V_{CD}/I_{AB} \qquad (4.20)$$

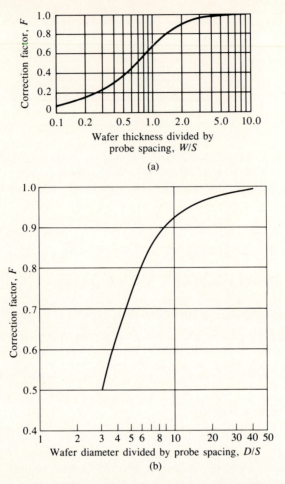

Fig. 4.18 Four-point-probe correction factors, F, used to correct for (a) wafers which are relatively thick compared to the probe spacing s and (b) wafers of finite diameter. In each case, $\rho = F\rho_{\text{measured}}$. (a) Copyright 1975 by McGraw-Hill Book Company. Reprinted with permission from ref. [12]. (b) Reprinted from ref. [27] with permission from the AT&T Technical Journal. Copyright 1958 AT&T.

Specially designed sheet-resistance test structures are often included on wafers so that the sheet resistances of n-type and p-type diffusions can be measured after final processing of the wafer. A sample structure is shown in Fig. 4.19.

4.8 CONCENTRATION-DEPENDENT DIFFUSION

Diffusion follows the theory of Section 4.3 as long as the impurity concentration remains below the value of the intrinsic-carrier concentration n_i at the diffusion temperature.

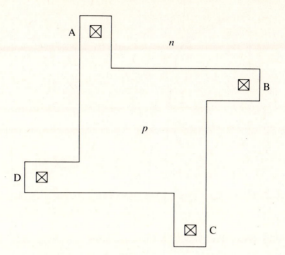

Fig. 4.19 A simple van der Pauw test structure used to measure the sheet resistance of a diffused layer. Sheet resistance is calculated using eq. (4.20).

Above this concentration, the diffusion coefficient becomes concentration-dependent. Each of the common impurities exhibits a different behavior.

The diffusion equation can be solved analytically for linear, parabolic, and cubic dependencies of the diffusion coefficient on concentration. The results are presented in Fig. 4.20, in which D_{sur} represents the diffusion coefficient at the surface. In general, concentration-dependent diffusion results in a much more abrupt profile than for the case of a constant-diffusion coefficient.

Boron and arsenic can be modeled by the first-order dependence in Fig. 4.20, resulting in the analytical relations between junction depth, sheet resistance, total dose, and surface concentration given in Table 4.2.[7-9]

High-concentration phosphorus diffusion results in a more complicated profile than that of boron or arsenic. Figure 4.21 depicts typical shallow phosphorus diffusion profiles. As phosphorus diffuses into the wafer, the diffusion coefficient becomes enhanced at concentrations below approximately $10^{19}/cm^3$, resulting in a distinct "kink" in the profile. The kink effect represents a practical limitation to the use of phosphorus for the

Table 4.2 Properties of High-Concentration Arsenic and Boron Diffusions.

Element	$x_j(cm)$	$D(cm^2/sec)$	$N_0(cm^{-3})$	$Q(cm^{-2})$
Arsenic	$2.29\sqrt{N_0 Dt/n_i}$*	$22.9\ exp(-4.1/kT)$	$1.56 \times 10^{17}(R_s x_j)^{-1}$	$0.55 N_0 x_j$
Boron	$2.45\sqrt{N_0 Dt/n_i}$*	$3.17\ exp(-3.59/kT)$	$2.78 \times 10^{17}(R_s x_j)^{-1}$	$0.67 N_0 x_j$

*The value of n_i must be calculated at the diffusion temperature.

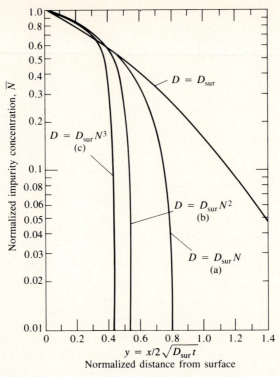

Fig. 4.20 Diffusion profiles for concentration-dependent diffusion. Copyright 1963 by the American Physical Society. Reprinted with permission from ref. [6].

source/drain diffusions and emitter diffusions of shallow MOS and bipolar devices. Most MOS and bipolar VLSI processes now use arsenic to avoid this problem. Complex mathematical models describing the diffusion of phosphorus may be found in refs. [10] and [11].

4.9 PROCESS SIMULATION

As the scale of integrated circuits is reduced, accurate knowledge of one-, two-, and even three-dimensional impurity profiles is becoming more and more important. At the same time, experimental determination of profiles is becoming a very difficult and time-consuming task in VLSI fabrication processes.

Sophisticated computer programs which can predict the results of fabrication steps have become available.[15–19] These programs not only numerically solve the generalized nonlinear diffusion equation in silicon, but also include the ability to simulate oxide growth with its attendant moving Si-SiO$_2$ boundary, impurity segregation during oxide growth, dopant evaporation from the surface, and ion implantation.

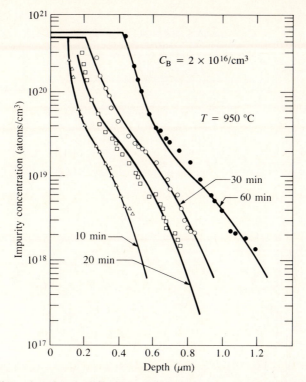

Fig. 4.21   Shallow phosphorus diffusion profiles for constant-source diffusions at 950 °C. Copyright 1969 IEEE. Reprinted with permission from ref. [10].

One of the most widely used of these programs is SUPREM, the *S*tanford *U*niversity *P*rocess *E*ngineering *M*odeling program.[15, 16, 19] The use of SUPREM requires specification of the process steps, including times, temperatures, and other ambient conditions for oxidation, diffusion, ion implantation, film deposition, and etching. The program calculates the impurity profile in the silicon substrate as well as in oxide and polysilicon layers. (See Problem 4.15.)

Simulation is growing in importance throughout the VLSI fabrication process. The detailed structures of recessed oxidation are being simulated, as are the photoresist and etching profiles resulting from processing at submicron dimensions. Ref. [20] contains a recent overview of process simulation.

4.10   DIFFUSION SYSTEMS

The open-furnace-tube system using solid, liquid, or gaseous sources, as depicted in Fig. 4.22, yields good reproducibility and is the most common diffusion technology used in integrated-circuit production today. The furnace tubes are the same as those used in

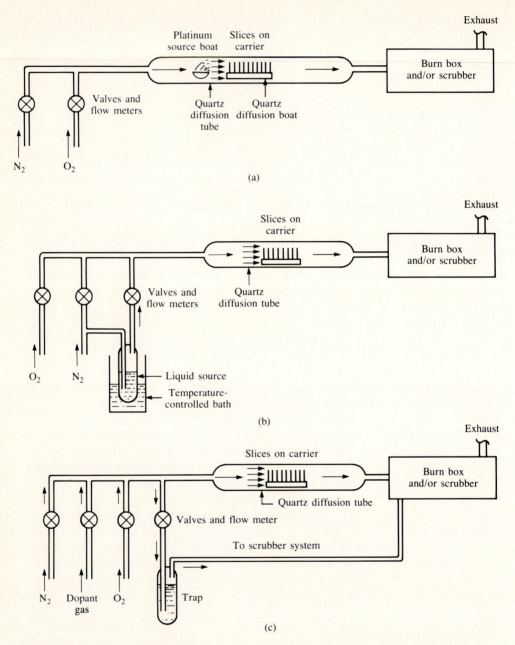

Fig. 4.22 Open-furnace-tube diffusion systems. (a) Solid source in a platinum source boat in the rear of diffusion tube; (b) liquid-source system with carrier gas passing through a bubbler; (c) diffusion system using gaseous impurity sources. Copyright John Wiley and Sons. Reprinted with permission from Ref. [23].

the three-zone systems described in Chapter 3. For diffusion, wafers are placed in a quartz boat and positioned in the center zone of the furnace, where the wafers are heated to a high temperature. Impurities are transported to the silicon surface, where they diffuse into the wafer.

Most common silicon dopants can be applied using liquid spin-on sources. These spin-on dopants are versatile, safe, and easy to apply, but the uniformity is often poorer than with other impurity sources. To achieve good quality control, most production systems use other solid, liquid, or gaseous impurity sources.

In one type of solid-source system, carrier gases (usually N_2 or O_2) flow at a controlled rate over a source boat placed in the furnace tube. The carrier gas picks up the vapor from the source and transports it down the tube, where the dopant species is deposited on the surface of the wafer. The temperature of the source is controlled to maintain the desired vapor pressure. The source can be placed in a low-temperature section of the furnace or may be external to the furnace. Solid boron and phosphorus impurity sources are also available in wafer form and are placed in the boat between adjacent pairs of silicon wafers.

In liquid-source systems, a carrier gas passes through a bubbler where it picks up the vapor of the liquid source. The gas carries the vapor into the furnace tube where it reacts with the surface of the silicon wafer.

Gas-source systems supply the dopant species directly to the furnace tube in the gaseous state. The common gas sources are extremely toxic, and additional input purging and trapping systems are required to ensure that all the source gas is removed from the system before wafer entry or removal. In addition, most diffusion processes either do not use all of the source gas or produce undesirable reaction by-products. Therefore, the output of diffusion systems should be processed by burning or by chemical and/or water scrubbing before being exhausted into the atmosphere.

Boron is the only commonly used p-type dopant. The diffusion coefficients of aluminum and gallium are quite high in silicon dioxide, and these elements cannot be masked effectively by SiO_2. Indium is not used because it is a relatively deep-level acceptor ($E_A - E_V = 0.14$ eV).

In contrast, antimony, phosphorus, and arsenic can all be masked by silicon dioxide and are all routinely used as n-type dopants in silicon processing.

4.10.1 Boron Diffusion

Boron has a high solubility in silicon and can achieve active surface concentrations as high as $4 \times 10^{20}/cm^3$ (Fig. 4.6). Elemental boron is inert up to temperatures exceeding the melting point of silicon. A surface reaction with boron trioxide (B_2O_3) is used to introduce boron to the silicon surface:

$$2B_2O_3 + 3Si \rightleftharpoons 4B + 3SiO_2 \tag{4.21}$$

An excess amount of boron trioxide can cause formation of a brown boron skin which is very difficult to remove with most acids. Boron skin formation can be minimized by performing the diffusions in an oxidizing atmosphere containing 3 to 10% oxygen. In a two-step diffusion, the boron predeposition step is commonly followed by a short wet-oxidation step to assist in removal of the boron skin prior to drive-in.

Common solid sources of boron include trimethylborate (TMB) and boron nitride wafers. TMB is a solid with high vapor pressure at room temperature. The TMB source is normally placed outside the diffusion furnace and cooled below room temperature during use. TMB vapor reacts in the furnace tube with oxygen to form boron trioxide, water, and carbon dioxide:

$$2(CH_3O)_3B + 9O_2 \xrightarrow{900\ °C} B_2O_3 + 6CO_2 + 9H_2O \qquad (4.22)$$

Any unreacted TMB should be scrubbed from the exhaust stream.

Boron nitride is a solid source available in wafer form. Activated wafers are placed in every third slot in the same quartz boat used to hold the silicon wafers. A silicon wafer faces each side of the oxidized boron nitride wafer, and boron trioxide is transferred directly to the surface of the silicon wafer during high-temperature diffusion. A small flow of inert gas such as nitrogen is used to keep contaminants out of the tube during diffusion.

The most common liquid source for boron is boron tribromide (BBr_3). The reaction is

$$4BBr_3 + 3O_2 \longrightarrow 2B_2O_3 + 6Br_2 \qquad (4.23)$$

Free bromine combines easily with metallic impurities and is useful in removing (gettering) metallic impurities during diffusion. Bromine, as well as unused boron tribromide, is in the exhaust stream, so the outlet gases should be carefully cleaned.

The primary gaseous source of boron is diborane (B_2H_6). Diborane is a highly poisonous and explosive gas. Table 4.3 summarizes the ACGIH recommendations for the maximum permissible exposure to the common gases used as diffusion sources. Extreme care must be taken in using these gases. In order to reduce the risk of handling, diborane is usually diluted with 99.9% argon or nitrogen by volume.

Diborane oxidizes in either oxygen or carbon dioxide to form boron trioxide:

$$B_2H_6 + 3O_2 \xrightarrow{300\ °C} B_2O_3 + 3H_2O \qquad (4.24)$$

and

$$B_2H_6 + 6CO_2 \longrightarrow B_2O_3 + 6CO + 3H_2O \qquad (4.25)$$

Both systems must provide a means for purging diborane from the input to the diffusion tube, and the output must be scrubbed to eliminate residual diborane and carbon monoxide.

Table 4.3 Threshold Limit Recommendations for Common Gaseous Sources.*[21]

Source	8-h exposure level (ppm)	Life-threatening exposure	Comments
Diborane (B_2H_6)	0.10	160 ppm for 15 min	Colorless, sickly sweet, extremely toxic, flammable.
Phosphine (PH_3)	0.30	400 ppm for 30 min	Colorless, decaying fish odor, extremely toxic, flammable. A few minutes' exposure to 2000 ppm can be lethal.
Arsine (AsH_3)	0.05	6–15 ppm for 30 min	Colorless, garlic odor, extremely toxic. A few minutes' exposure to 500 ppm can be lethal.
Silane (SiH_4)	0.50	Unknown	Repulsive odor, burns in air, explosive, poorly understood.
Dichlorosilane (SiH_2Cl_2)	5.00	. . .	Colorless, flammable, toxic. Irritating odor provides adequate warning for voluntary withdrawal from contaminated areas.

*Data from the 1979 American Conference of Governmental Hygienists (ACGIH).

4.10.2 Phosphorus Diffusion

Phosphorus has a higher solubility in silicon than does boron, and surface concentrations in the low $10^{21}/cm^3$ range can be achieved during high-temperature diffusion. Phosphorus is introduced into silicon through the reaction of phosphorus pentoxide at the wafer surface:

$$2P_2O_5 + 5Si \rightleftharpoons 4P + 5SiO_2 \tag{4.26}$$

Solid P_2O_5 wafers can be used as a solid source for phosphorus, as can ammonium monophosphate ($NH_4H_2PO_4$) and ammonium diphosphate [$(NH_4)_2H_2PO_4$] in wafer form. However, the most popular diffusion systems use either liquid or gaseous sources. Phosphorus oxychloride ($POCl_3$) is a liquid at room temperature. A carrier gas is passed through a bubbler and brings the vapor into the diffusion furnace. The gas stream also contains oxygen, and P_2O_5 is deposited on the surface of the wafers:

$$4POCl_3 + 3O_2 \longrightarrow 2P_2O_5 + 6Cl_2 \tag{4.27}$$

Liberated chlorine gas serves as a gettering agent, and Cl_2 and $POCl_3$ must be removed from the exhaust stream.

Phosphine, PH_3, is a highly toxic and explosive gas used as the gaseous source for phosphorus. It is also supplied in dilute form with 99.9% argon or nitrogen. Phosphine is oxidized with oxygen in the furnace:

$$2PH_3 + 4O_2 \longrightarrow P_2O_5 + 3H_2O \qquad (4.28)$$

Unreacted phosphine must be cleaned from the exhaust gases, and the gas delivery system must be able to purge phosphine from the input to the tube.

4.10.3   Arsenic Diffusion

Arsenic has the highest solubility of any of the common dopants in silicon, with surface concentrations reaching $2 \times 10^{21}/cm^3$. The surface reaction involves arsenic trioxide:

$$2As_2O_3 + 3Si \rightleftharpoons 3SiO_2 + 4As \qquad (4.29)$$

Oxide vapors can be carried into the furnace tube from a solid diffusion source by a nitrogen carrier gas. However, evaporation of arsenic from the surface limits surface concentrations to below $3 \times 10^{19}/cm^3$. The exhaust must be carefully cleaned because of the presence of arsenic.

Arsine gas may be used as a source, but it is extremely toxic and also produces relatively low surface concentrations. The problems with arsenic deposition and safety delayed its widespread use in silicon processing until ion implantation was developed in the early 1970s. Ion implantation is now the preferred technique for introducing arsenic into silicon. (Chapter 5 is devoted to the subject of ion implantation.)

4.10.4   Antimony Diffusion

Antimony, like arsenic, has a low diffusion coefficient and has been used for a long time for buried layers in bipolar processes. Antimony trioxide is a solid source which is placed in a two-zone furnace in which the source is maintained at a temperature of 600 to 650 °C. Antimony is introduced at the silicon surface as in the other cases:

$$2Sb_2O_3 + 3Si \rightleftharpoons 3SiO_2 + 4Sb \qquad (4.30)$$

A liquid source, antimony pentachloride (Sb_3Cl_5), has been successfully used with oxygen as a carrier gas passing through a bubbler. The gas stabine (SbH_3) is unstable and cannot be used for antimony diffusion.

4.11   SUMMARY

In Chapter 4 we have discussed the formation of *pn* junctions using high-temperature diffusion. Mathematical models for diffusion have been presented, and the behavior of common *n*- and *p*-type dopants in silicon has been discussed. A key parameter governing the diffusion process is the diffusion coefficient, which is highly temperature-dependent, following an Arrhenius relationship. Boron, phosphorus, antimony, and arsenic all have reasonable diffusion coefficients in silicon at temperatures between 900 and 1200 °C, and they can be conveniently masked by a barrier layer of silicon dioxide. Gallium and aluminum are not easily masked by SiO_2 and are seldom used, and indium is not used because of its large activation energy. At high concentrations, diffusion coefficients become concentration-dependent, causing diffused profiles to differ substantially from predictions of simple theories.

Two types of diffusions are most often used. If the surface concentration is maintained constant throughout the diffusion process, then a complementary error function (erfc) distribution is obtained. In the erfc case, the surface concentration is usually set by the solid-solubility limit of the impurity in silicon. If a fixed dose of impurity is diffused into silicon, a Gaussian diffusion profile is achieved. These two cases are often combined in a two-step process to obtain lower surface concentrations than those achievable with a solid-solubility-limited diffusion. As the complexity of fabrication processes grows, simulation with process modeling programs such as SUPREM is becoming ever more important.

The concept of sheet resistance has been introduced, and Irvin's curves have been used to relate the sheet resistance, junction depth, and surface concentration of diffused layers. Techniques for calculating and measuring junction depth have also been presented. Resistor fabrication has been discussed, including end and corner effects, as well as the effects of lateral diffusion under the edges of diffusion barriers.

High-temperature open-furnace diffusion systems are routinely used for diffusion with solid, liquid, and gaseous impurity sources. Boron, phosphorus, and antimony are all easily introduced into silicon using high-temperature diffusion. However, arsenic deposition by diffusion is much more difficult, and today it is usually accomplished using ion implantation (see Chapter 5). As with many chemicals used in integrated-circuit fabrication, some of the sources used for diffusion are extremely toxic and must be handled with great care.

REFERENCES

[1] R. F. Pierret, *Semiconductor Fundamentals,* Volume I in the Modular Series on Solid State Devices, Addison-Wesley, Reading, MA, 1983.

[2] R. A. Colclaser, *Microelectronics: Processing and Device Design,* John Wiley & Sons, New York, 1980.

[3] R. F. Pierret, *Advanced Semiconductor Fundamentals,* Volume VI in the Modular Series on Solid State Devices, Addison-Wesley, Reading, MA, 1987.

[4] D. P. Kennedy and R. R. O'Brien, "Analysis of the Impurity Atom Distribution Near the Diffusion Mask for a Planar p-n Junction," IBM Journal of Research & Development, *9,* 179–186 (May, 1965).

[5] J. C. Irvin, "Resistivity of Bulk Silicon and of Diffused Layers in Silicon," Bell System Technical Journal, *41*, 387–410 (March, 1962).

[6] L. R. Weisberg and J. Blanc, "Diffusion with Interstitial-Substitutional Equilibrium: Zinc in GaAs," Physical Review, *131*, 1548–1552 (August 15, 1963).

[7] R. B. Fair, "Boron Diffusion in Silicon — Concentration and Orientation Dependence, Background Effects, and Profile Estimation," Journal of the Electrochemical Society, *122*, 800–805 (June, 1975).

[8] R. B. Fair, "Profile Estimation of High-Concentration Arsenic Diffusions in Silicon," Journal of Applied Physics, *43*, 1278–1280 (March, 1972).

[9] R. B. Fair and J. C. C. Tsai, "Profile Parameters of Implanted-Diffused Arsenic Layers in Silicon," Journal of the Electrochemical Society, *123*, 583–586 (1976).

[10] J. C. C. Tsai, "Shallow Phosphorus Diffusion Profiles in Silicon," Proceedings of the IEEE, *57*, 1499–1506 (September, 1969).

[11] R. B. Fair and J. C. C. Tsai, "A Quantitative Model for the Diffusion of Phosphorus in Silicon and the Emitter Dip Effect," Journal of the Electrochemical Society, *124*, 1107–1118 (July, 1977).

[12] W. R. Runyan, *Semiconductor Measurements and Instrumentation,* McGraw-Hill, New York, 1975.

[13] L. J. van der Pauw, "A Method of Measuring Specific Resistivity and Hall Effect of Discs of Arbitrary Shape," Philips Research Reports, *13*, 1–9 (February, 1958).

[14] R. Chwang, B. J. Smith, and C. R. Crowell, "Contact Size Effects on the van der Pauw Method for Resistivity and Hall Coefficient Measurements," Solid-State Electronics, *17*, 1217–1227 (December, 1974).

[15] D. A. Antoniadis and R. W. Dutton, "Models for Computer Simulation of Complete IC Fabrication Processes," IEEE Journal of Solid State Circuits, *SC-14*, 412–422 (April, 1979).

[16] C. P. Ho, J. D. Plummer, S. E. Hansen, and R. W. Dutton, "VLSI Process Modeling — SUPREM III," IEEE Trans. Electron Devices, *ED-30*, 1438–1453 (November, 1983).

[17] D. Chin, M. Kump, H. G. Lee, and R. W. Dutton, "Process Design Using Coupled 2D Process and Device Simulators," IEEE IEDM Digest, p. 223–226 (December, 1980).

[18] C. D. Maldanado, F. Z. Custode, S. A. Louie, and R. K. Pancholy, "Two Dimensional Simulation of a 2 μm CMOS Process Using ROMANS II," IEEE Trans. Electron Devices, *30*, 1462–1469 (November, 1983).

[19] M. E. Law, C. S. Rafferty, and R. W. Dutton, "New *n*-well Fabrication Techniques Based on 2D Process Simulation," IEEE IEDM Digest, p. 518–521 (December, 1986).

[20] R. W. Dutton, "Modeling and Simulation for VLSI," IEEE IEDM Digest, p. 2–7 (December, 1986).

[21] *Matheson Gas Data Book,* Matheson Gas Products, 1980.

[22] A. B. Glaser and G. E. Subak-Sharpe, *Integrated Circuit Engineering,* Addison-Wesley, Reading, MA, 1979.

[23] S. K. Ghandhi, *VLSI Fabrication Principles,* John Wiley & Sons, New York, 1983.

[24] S. M. Sze, Ed., *VLSI Technology,* McGraw-Hill, New York, 1983.

[25] S. K. Ghandhi, *The Theory and Practice of Microelectronics,* John Wiley & Sons, New York, 1968.

[26] R. B. Fair, "Recent Advances in Implantation and Diffusion Modeling for the Design and Process Control of Bipolar ICs," Semiconductor Silicon 1977, *PV 77-2,* p. 968–985.

[27] F. M. Smits, "Measurement of Sheet Resistivities with the Four Point Probe," Bell System Technical Journal, *37,* 711–718 (May, 1958).

PROBLEMS

4.1 A phosphorus diffusion has a surface concentration of $5 \times 10^{18}/cm^3$, and the background concentration of the p-type wafer is $1 \times 10^{15}/cm^3$. The Dt product for the diffusion is 10^{-8} cm^2.

(a) Find the junction depth for a Gaussian distribution.

(b) Find the junction depth for an erfc profile.

(c) What is the sheet resistance of the two diffusions?

(d) Draw a graph of the two profiles.

4.2 A 5-hr boron diffusion is to be performed at 1100 °C.

(a) What thickness of silicon dioxide is required to mask this diffusion?

(b) Repeat part (a) for phosphorus.

4.3 A boron diffusion into a 1-ohm-cm n-type wafer results in a Gaussian profile with a surface concentration of $5 \times 10^{18}/cm^3$ and a junction depth of 4 μm.

(a) How long did the diffusion take if the diffusion temperature was 1100 °C?

(b) What was the sheet resistance of the layer?

(c) What is the dose in the layer?

(d) The boron dose was deposited by a solid-solubility-limited diffusion. Design a diffusion schedule (temperature and time) for this predeposition step.

4.4 The boron diffusion in Problem 4.3 is followed by a solid-solubility-limited phosphorus diffusion for 30 min at 950 °C. Assume that the boron profile does not change during the phosphorus diffusion.

(a) Find the junction depth of the new phosphorus layer. Assume an erfc profile.

(b) Find the junction depth based on the concentration-dependent diffusion data presented in Fig. 4.21.

4.5 The p-well in a CMOS process is to be formed by a two-step boron diffusion into a 5-ohm-cm n-type substrate. The sheet resistance of the well is .1000 ohms per square and the junction depth is 7.5 μm.

(a) Design a reasonable diffusion schedule for the drive-in step which produces the p-well.

(b) What is the final surface concentration in the p-well?

(c) What is the dose required to form the well?

(d) Can this dose be achieved using a solid-solubility-limited diffusion with diffusion temperatures of 900 °C or above? Discuss.

4.6 The channel length of a metal-gate NMOS transistor is the spacing between the source and drain diffusions as shown in Fig. P4.6a. The spacing between the source and drain diffusion openings is 6 μm on the masking oxide used to make the transistor. The source/drain junctions are diffused to a depth of 1.5 μm using a constant-source diffusion. The surface concentration is $1 \times 10^{20}/cm^3$ and the wafer has a concentration of $1 \times 10^{16}/cm^3$. What is the channel length in the actual device after the diffusion is completed?

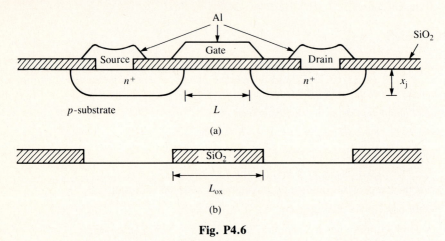

(a)

(b)

Fig. P4.6

4.7 (a) What is the total number of squares in the resistor shown in Fig. P4.7 assuming that its geometry is specified precisely by the mask dimensions?

(b) The resistor is actually formed from a p-type base diffusion with a 6-μm junction depth. What is the actual number of squares in this resistor, assuming that the lateral diffusion under the edge of the mask is 5 μm.

(c) What would be the resistance of the resistors in parts (a) and (b) if the surface concentration of the base diffusion was 5×10^{18} boron atoms/cm^3, the bulk concentration $10^{15}/cm^3$, and the junction depth 6 μm.

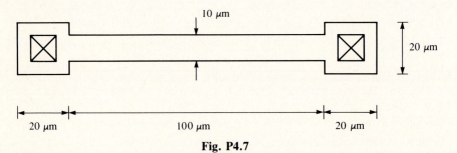

Fig. P4.7

4.8 In practice, wafers are slowly pushed into and pulled out of the furnace, or the furnace temperature may be changed with time. Assume that the furnace temperature is being ramped down with time: $T = T_0 - Rt$, where T_0 is the initial temperature and R is the temperature change per second. Show that the effective Dt product defined by

$$(Dt)_{\text{eff}} = \int_0^{t_0} D(t)\,dt$$

where t_0 is the ramp-down time is given by

$$(Dt)_{\text{eff}} = D(T_0)\,(kT_0^2/RE_A)$$

where

$$D(T_0) = D_0\,\exp(-E_A/kT_0)$$

4.9 Determine the sensitivity of junction depth to changes in furnace temperature by calculating $(dx_j/x_j)/(dT/T)$ for a Gaussian diffusion profile. What fractional change in junction depth will occur at 1100 °C if the furnace temperature is in error by 10 °C?

4.10 Rework Example 4.2 using the concentration-dependent boron diffusion expressions for the predeposition calculations. Find the new surface concentration and junction depth following the drive-in step and compare the results with those presented in the example. At 900 °C, $n_i = 4 \times 10^{18}/\text{cm}^3$.

4.11 Derive the expressions for the Gaussian and complementary-error-function solutions to the diffusion equation.

4.12 What is the minimum sheet resistance to be expected from shallow arsenic- and boron-doped regions if the regions are 1 μm deep? 0.25 μm deep? Make use of Fig. 4.6 and your knowledge of the dependence of mobility on doping (Volume I). Assume that the region is uniformly doped. Compare your results to the equations presented in Section 4.8.

4.13 Gold is diffused into a silicon wafer using a constant-source diffusion with a surface concentration of $10^{18}/\text{cm}^3$. How long does it take the gold to diffuse completely through a silicon wafer 400 μm thick with a background concentration of $10^{16}/\text{cm}^3$ at a temperature of 1000 °C?

4.14 A gas cylinder contains 100 ft^3 of a mixture of diborane and argon. The diborane represents 0.1% by volume. An accident occurs and the complete cylinder is released into a room measuring $10 \times 12 \times 8$ ft.

(a) What will be the equilibrium concentration of diborane in the room in ppm?

(b) Compare this level with the toxic level based on Table 4.3.

(c) Would your answer to part (b) change if the gas cylinder contained arsine?

4.15 (a) Use SUPREM to simulate the diffusion profile of Example 4.2. Compare the simulation results with those given in the example.

(b) Follow the boron diffusion by the growth of a 500-nm layer of oxide in wet oxygen at 1100 °C. Discuss what has happened to the boron concentration at the Si-SiO$_2$ interface.

(c) Add a 30-min solid-solubility-limited phosphorus diffusion at 1000 °C.

(d) The phosphorus diffusion created a new pn junction. Update the hand calculations for the boron impurity profile of Example 4.2 and estimate the location of both pn junctions with the aid of Fig. 4.21. Compare your results to those of SUPREM in part (c).

5 / Ion Implantation

Ion implantation offers many advantages over diffusion for the introduction of impurity atoms into the silicon wafer and has become a workhorse technology in modern integrated-circuit fabrication. In this chapter we will first discuss ion implantation technology and mathematical modeling of the impurity distributions obtained with ion implantation. We will subsequently explore deviations from the model caused by nonideal behavior and will discuss annealing techniques used to remove crystal damage caused by the implantation process.

5.1 IMPLANTATION TECHNOLOGY

An ion implanter is a high-voltage particle accelerator producing a high-velocity beam of impurity ions which can penetrate the surface of silicon target wafers. The basic parts of the system, shown schematically in Fig. 5.1, are described in detail below, beginning with the impurity-source end of the system.

1. **Ion Source.** The ion source operates at a high voltage (25 kV) and produces a plasma containing the desired impurity as well as other undesired species. Arsine, phosphine, and diborane, as well as other gases, can be used in the source. Solids can be sputtered in special ion sources, and this technique offers a wide degree of flexibility in the choice of impurity.

2. **Mass Spectrometer.** An analyzer magnet bends the ion beam through a right angle to select the desired impurity ion. The selected ion passes through an aperture slit into the main accelerator column.

3. **High-Voltage Accelerator.** The accelerator column adds energy to the beam (up to 175 keV) and accelerates the ions to their final velocity. Both the accelerator column and the ion source are operated at a high voltage relative to the target. For protection from high voltage and possible X-ray emission, the ion source and accelerator are mounted within a protective shield.

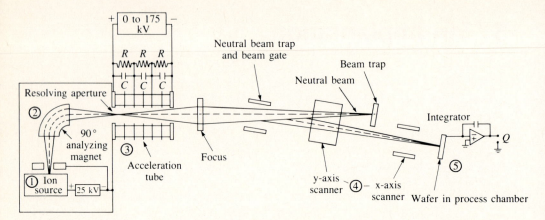

Fig. 5.1   Schematic drawing of a typical ion implanter showing (1) ion source, (2) mass spectrometer, (3) high-voltage accelerator column, (4) x- and y-axis deflection system, and (5) target chamber.

4. **Scanning System.** X- and y-axis deflection plates are used to scan the beam across the wafer to give a uniform implantation and to build up the desired dose. The beam is bent slightly to prevent neutral particles from hitting the target.

5. **Target Chamber.** Silicon wafers serve as targets for the ion beam. For safety, the target area is maintained near ground potential. The complete implanter system is operated under vacuum conditions.

The analyzer magnet is used to select the desired impurity ions from the output of the source. A charged particle moving with velocity $\bar{v}$ through a magnetic field $\bar{B}$ will experience a force $\bar{F}$, given by

$$\bar{F} = q(\bar{v} \times \bar{B}) \tag{5.1}$$

The force will tend to move the particle in a circle, and the centrifugal force will balance $\bar{F}$. For the case where $\bar{B}$ is perpendicular to $\bar{v}$, $q|\bar{v}||\bar{B}| = m|\bar{v}|^2/r$ where $m|\bar{v}|^2/2 = qV$ and V is the accelerator voltage. Thus the magnitude of the magnetic field $\bar{B}$ may be adjusted to select an ion species with a given mass:

$$|\bar{B}| = \sqrt{(2mV/qr^2)} \tag{5.2}$$

The ion source in Fig. 5.1 operates at a constant potential (25 keV) so that the voltage V is known, and an ion species is selected by changing the dc current supplying the analyzer magnet. The selected impurity is then accelerated to its final velocity in the high-voltage column.

The silicon wafer is maintained in good electrical contact with the target holder, so electrons can readily flow to or from the wafer to neutralize the implanted ions. This electron current is integrated over time to measure the total dose Q from the implanter given by

$$Q = \int_0^T I\,dt/nqA \tag{5.3}$$

where I is the beam current in amperes, A is the wafer area, $n = 1$ for singly ionized ions and 2 for doubly ionized species, and T is the implantation time. The use of a doubly ionized species increases the energy capability of the machine by a factor of 2 since $E = nqV$.

The target wafers can be maintained at relatively low temperatures during the implantation. Low-temperature processing prevents undesired spreading of impurities by diffusion, which is very important in VLSI fabrication. Another advantage of ion implantation is the ability to use a much wider range of impurity species than possible with diffusion. In principle, any element that can be ionized can be introduced into the wafer using implantation.

A production-level ion implanter may cost from \$1 to \$2 million, and cost is its greatest disadvantage. However, the advantages of flexibility and process control have far outweighed the disadvantage of cost, and ion implantation is now used routinely throughout bipolar and MOS integrated-circuit fabrication.

5.2   MATHEMATICAL MODEL FOR ION IMPLANTATION

As an ion enters the surface of the wafer, it collides with atoms in the lattice and interacts with electrons in the crystal. Each nuclear or electronic interaction reduces the energy of the ion until it finally comes to rest within the target. Interaction with the crystal is a statistical process, and the implanted impurity profile can be approximated by the Gaussian distribution function illustrated in Fig. 5.2. The distribution is described mathematically by

$$N(x) = N_p \exp[-(x - R_p)^2/2\,\Delta R_p^2] \tag{5.4}$$

R_p is called the *projected range* and is equal to the average distance an ion travels before it stops. The peak concentration N_p occurs at $x = R_p$. Because of the statistical nature of the process, some ions will be "lucky" and will penetrate beyond the projected range R_p, and some will be "unlucky" and will not make it as far as R_p. The spread of the distribution is characterized by the standard deviation, ΔR_p, called the *straggle*.

The area under the impurity distribution curve is the implanted dose Q, defined by

$$Q = \int_0^\infty N(x)\,dx \tag{5.5}$$

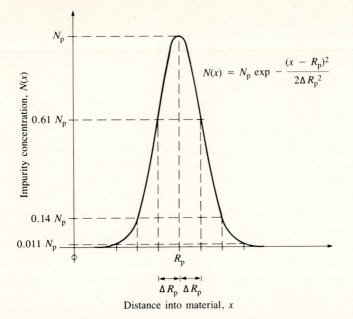

Fig. 5.2 Gaussian distribution resulting from ion implantation. The impurity is shown implanted completely below the wafer surface ($x = 0$).

For an implant completely contained within the silicon, the dose is equal to

$$Q = \sqrt{2\pi}\,N_p\,\Delta R_p \tag{5.6}$$

Implanted doses can range from $10^{10}/cm^2$ to $10^{18}/cm^2$, and ion implantation is often used to replace the predeposition step in a two-step diffusion process. Doses in the range of 10^{10} to $10^{13}/cm^2$ are required for threshold adjustment in MOS technologies and are almost impossible to achieve using diffusion. CMOS well formation is another example where the dose control of the ion implanter is a distinct advantage. Doses exceeding $10^{15}/cm^2$ are quite large and can be quite time-consuming to produce using ion implantation. As a reference for comparison, the silicon lattice atomic sheet density is approximately 7×10^{14} silicon atoms/cm^2 on the $\langle 100 \rangle$ surface.

The implanted dose can be controlled within a few percent, and this tight control represents another advantage of ion implantation. For example, resistors can be fabricated with absolute tolerances of a few percent in carefully controlled processes using ion implantation, whereas the same resistors would have an absolute tolerance exceeding 20% if they were formed using only diffusion.

The projected range of a given ion is a function of the energy of the ion, and of the mass and atomic number of both the ion and the target material. A theory for range and straggle was developed by Lindhard, Scharff, and Schiott and is called the *LSS theory*.[1]

This theory assumes that the implantation goes into an amorphous material in which the atoms of the target material are randomly positioned. Figure 5.3 displays the results of LSS calculations for the projected range and straggle for antimony, boron, phosphorus, and arsenic in amorphous silicon and silicon dioxide. For the moment we will assume that these results are also valid for crystalline silicon. Deviations from the LSS theory will be discussed in Section 5.5.

Range and straggle are roughly proportional to ion energy over a wide range, although some nonlinear behavior is clearly evident in Fig. 5.3. For a given energy, the lighter elements strike the silicon wafer with a higher velocity and penetrate more deeply into the wafer. The results indicate that the projected ranges in Si and SiO_2 are essentially the same, and we will assume that the stopping power of silicon dioxide is equal to that of silicon. Figure 5.3 also gives values for the transverse straggle $\Delta R_\perp$, which will be discussed in the next section.

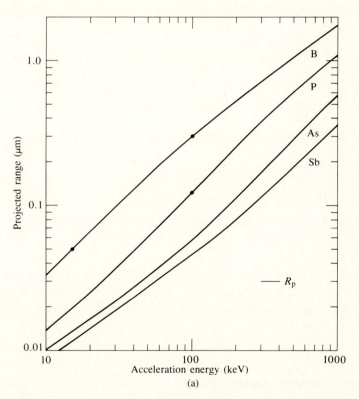

Fig. 5.3 Projected range and straggle calculations based on LSS theory. (a) Projected range R_p for boron, phosphorus, arsenic, and antimony in amorphous silicon. Results for SiO_2 and for silicon are virtually identical. (b) (On page 94) Vertical ΔR_p and transverse $\Delta R_\perp$ straggle for boron, phosphorus, arsenic, and antimony. Reprinted with permission from ref. [2]. (Copyright Van Nostrand Reinhold Company, Inc.)

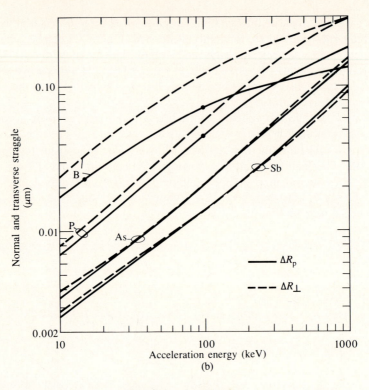

Fig. 5.3   (continued)

Example 5.1: Phosphorus with an energy of 100 keV is implanted into a silicon wafer. **(a)** What are the range and straggle associated with this implantation? **(b)** What should the implanted dose be if a peak concentration of $1 \times 10^{17}/cm^3$ is desired?

Solution: Using Fig. 5.3, the range and straggle are 0.12 μm and 0.045 μm, respectively. The dose and peak concentration are related by eq. (5.6). Note that this is an approximation, since the peak is only a little over $2\Delta R_p$ below the silicon surface.

$$Q = \sqrt{2\pi} N_p \Delta R_p = \sqrt{2\pi} \, (1 \times 10^{17}/cm^3)(4.5 \times 10^{-6} \, cm) = 1.13 \times 10^{12}/cm^2$$

5.3   SELECTIVE IMPLANTATION

In most cases we desire to implant impurities only in selected areas of the wafer. Windows are opened in a barrier material wherever impurity penetration is desired. In the center of the window the impurity distribution is described by eq. (5.4), but near the edges the distribution decreases and actually extends under the edge of the window, as shown in Fig. 5.4. The overall distribution can be modeled by[4]

$$N(x, y) = N(x)F(y)$$

$$F(y) = 0.5[\text{erfc}\{(y - a)/\sqrt{2}\,\Delta R_\perp\} - \text{erfc}\{(y + a)/\sqrt{2}\,\Delta R_\perp\}] \tag{5.7}$$

where $N(x)$ is given by eq. (5.4). The parameter $\Delta R_\perp$ is called the *transverse straggle* and characterizes the behavior of the distribution near the edge of the window. Figure 5.4 shows normalized impurity distributions near the barrier edge calculated with eq. (5.4). Figure 5.3b gives values of both normal straggle ΔR_p and transverse straggle $\Delta R_\perp$.

In order to mask the ion implantation, it is necessary to prevent the implanted impurity from changing the doping level in the silicon beneath the barrier layer. Figure 5.5 shows a silicon wafer with a layer of silicon dioxide on the surface. An

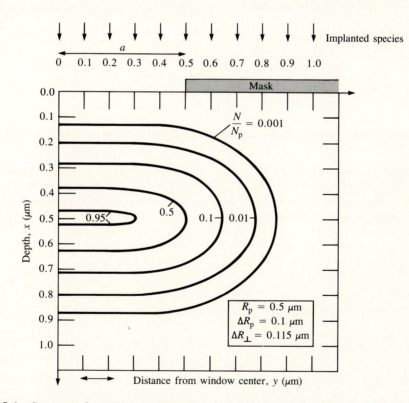

Fig. 5.4 Contours of equal ion concentration for an implantation into silicon through a 1-μm window. The profiles are symmetrical about the x-axis and were calculated using eq. (5.7), which is taken from ref. [4].

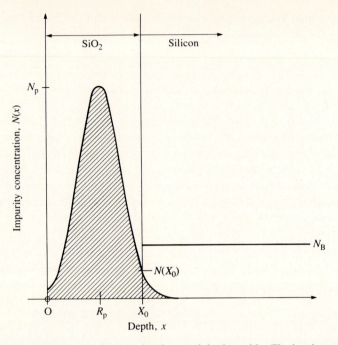

Fig. 5.5   Implanted impurity profile with implant peak in the oxide. The barrier material must be thick enough to ensure that the concentration in the tail of the distribution is much less than N_B.

impurity has been implanted into the wafer with the peak of the distribution in the silicon dioxide. To prevent significantly altering the doping in the silicon, we require that the implanted concentration be less than 1/10 the background concentration at the interface between the silicon and silicon dioxide:

$$N(X_0) < N_B/10$$

or

$$N_p \exp[-(X_0 - R_p)^2/2 \, \Delta R_p^2] < N_B/10 \tag{5.8}$$

Solving eq. (5.8) for X_0 yields a minimum oxide thickness of

$$X_0 = R_p + \Delta R_p \sqrt{2 \, \ln(10N_p/N_B)} = R_p + m \, \Delta R_p \tag{5.9}$$

The oxide thickness must be at least equal to the projected range plus some multiple m times the straggle. Table 5.1 gives values of m for various ratios of peak concentration to background concentration. An oxide thickness equal to the projected range plus six times the straggle should mask most ion implantations.

Table 5.1 Values of m for Various Values of N_p/N_B.

N_p/N_B	m
10^1	3.0
10^2	3.7
10^3	4.3
10^4	4.8
10^5	5.3
10^6	5.7

Silicon dioxide and silicon nitride are routinely used as barrier materials during implantation. Since implantation is a low-temperature process, additional materials such as photoresist and aluminum, which cannot withstand high-temperature diffusion, may be used as barrier materials during the implantation.

Silicon nitride is more effective than silicon dioxide in stopping ions, and a silicon nitride barrier layer need only be 85% of the thickness of an SiO_2 barrier layer. On the other hand, photoresist is less effective in stopping ions, and a photoresist barrier layer should be 1.8 times the thickness of an SiO_2 layer under the same implantation conditions. Metals are of such a high density that even a very thin layer will mask most implantations.

Example 5.2: A boron implantation is to be performed through a 50-nm gate oxide so that the peak of the distribution is at the Si-SiO_2 interface. The dose of the implant in silicon is to be 1×10^{13}/cm^2. **(a)** What are the energy of the implant and the peak concentration at the interface? **(b)** How thick should the SiO_2 layer be in areas which are not to be implanted, if the background concentration is 1×10^{16}/cm^3? **(c)** Suppose the oxide is 50 nm thick everywhere. How much photoresist is required on top of the oxide to completely mask the ion implantation?

Solution: The projected range needs to be 0.05 μm in order to place the peak of the distribution at the Si-SiO_2 interface. Using Fig. 5-3a, the R_p of 0.05 μm requires an energy of 15 keV. Since the peak of the implant is at the interface, the total dose will be twice the dose needed in silicon. The peak concentration is

$$N_p = Q/\Delta R_p\sqrt{2\pi} = 2 \times 10^{13}/(2.3 \times 10^{-6}\sqrt{2\pi}) = 3.5 \times 10^{18}/\text{cm}^3$$

where the straggle was found using Fig. 5-3b. In order to completely mask the implantation, the tail of the distribution must be less than the background concentration at the interface. The minimum oxide thickness is found using eq. (5.9):

$$X_0 = 0.05 + 0.023\sqrt{2 \ln(10 \times 3.5 \times 10^{18}/10^{16})} \ \mu\text{m} = 0.14 \ \mu\text{m}$$

Since the oxide is 0.05 μm thick, the photoresist must provide a thickness equivalent to 0.09 μm. The resist thickness must be 1.8 times the needed thickness of SiO_2 to provide an equivalent barrier layer, so the photoresist should be at least $0.16 \, \mu m$ thick. This thickness requirement is easily met with most photoresist layers.

5.4 JUNCTION DEPTH AND SHEET RESISTANCE

Ion implantation is often used to form shallow *pn* junctions for various device applications. The implanted profile approximates a Gaussian distribution, and the junction depth may be found by equating the implanted distribution to the background concentration, as in Chapter 4.

$$N_p \exp[-(x_j - R_p)^2/2 \, \Delta R_p^2] = N_B$$

$$x_j = R_p \pm \Delta R_p \sqrt{2 \ln(N_p/N_B)} \qquad (5.10)$$

Both roots may be meaningful, as indicated in Fig. 5.6, in which a deep subsurface implant has junctions occurring at two different depths, x_{j1} and x_{j2}.

Example 5.3: Boron is implanted into an *n*-type silicon wafer to a depth of $0.3 \, \mu m$. Find the location of the junction if the peak concentration is $1 \times 10^{18}/\text{cm}^3$ and the doping of the wafer is $3 \times 10^{16}/\text{cm}^3$.

Solution: From Fig. 5.3a, the implant energy is 100 keV. From Fig. 5.3b, the straggle is $0.07 \, \mu m$. Equating the Gaussian distribution to the background concentration yields

$$3 \times 10^{16} = 10^{18} \exp -(x_j - R_p)^2/2 \, \Delta R_p^2$$

or

$$x_j = R_p \pm 2.65 \, \Delta R_p$$

This yields junction depths of $0.12 \, \mu m$ and $0.49 \, \mu m$.

The peak of an implantation is often positioned at the silicon surface. For this special case, we may use Irvin's curves for Gaussian distributions to find the sheet resistance of the implanted layer, as discussed in Chapter 4. These curves may also be used to find the sheet resistance of a layer which is completely below the surface (see Problem 5.4). Note that an implanted Gaussian impurity distribution will remain Gaussian through any subsequent high-temperature processing steps.

Diffused profiles generally have the maximum impurity concentration at the silicon surface. Ion-implantation techniques can be used to produce profiles with subsurface peaks or "retrograde" profiles which decrease toward the wafer surface. Multiple implant steps at different energies can also be used to build up more complicated impurity profiles.

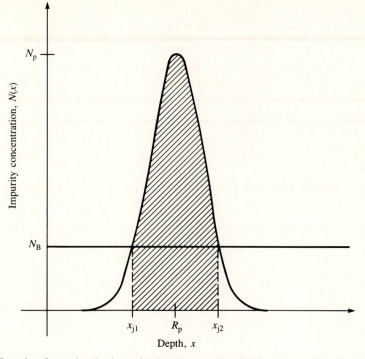

Fig. 5.6 Junction formation by impurity implantation in silicon. Two *pn* junctions are formed at x_{j1} and x_{j2}.

5.5 CHANNELING, LATTICE DAMAGE, AND ANNEALING

5.5.1 Channeling

The LSS results of Section 5.2 are based on the assumption that the target material is amorphous, having a completely random order. This assumption is true of thermal SiO_2, deposited Si_3N_4 and SiO_2, and many thin metal films, but it is not valid for a crystalline substrate. The regular arrangement of atoms in the crystal lattice leaves a large amount of open space in the crystal. Figure 5.7 shows a view through the silicon lattice in the $\langle 110 \rangle$ direction. If the incoming ion flux is improperly oriented with respect to the crystal planes, the ions will tend to miss the silicon atoms in the lattice and will "channel" much more deeply into the material than the LSS theory predicts. However, electronic inter-actions will eventually stop the ions.

The effects of channeling are demonstrated in Fig. 5.8. Phosphorus has been im-planted at an energy of 40 keV into a silicon target with several orientations of the ion beam relative to the $\langle 100 \rangle$ silicon surface. The appearance of a random target can be achieved by tilting $\langle 100 \rangle$ silicon approximately 7° relative to the incoming beam. The results are represented by the x's in Fig. 5.8. The range for this case compares well with

Fig. 5.7 The silicon lattice viewed along the ⟨110⟩ axis. From THE ARCHITECTURE OF MOLECULES by Linus Pauling and Roger Hayward. Copyright © 1964 W. H. Freeman and Company. Reprinted with permission from refs. [3a] and [3b].

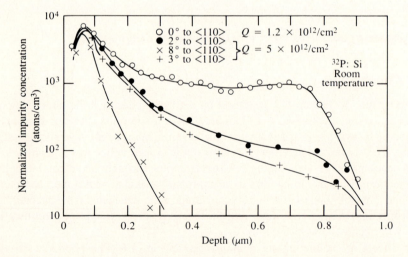

Fig. 5.8 Phosphorus impurity profiles for 40-keV implantations at various angles from the ⟨110⟩ axis. Copyright 1968 by National Research Council of Canada. Reprinted with permission from ref. [5].

the LSS calculations presented in Fig. 5.3. The open circles represent the boron profile implanted perpendicular to the ⟨100⟩ surface. Note that the range for the "channeled" case is almost twice that predicted by the LSS theory. Results for two other angles of incidence are given in Fig. 5.8, showing progressively less channeling as the angle is increased.

5.5.2   Lattice Damage and Annealing

During the implantation process, ion impact can knock atoms out of the silicon lattice, damaging the implanted region of the crystal. If the dose is high enough, the implanted layer will become amorphous. Figure 5.9 gives the dose required to produce an amorphous silicon layer for various impurities as a function of substrate temperature. The heavier the impurity, the lower the dose required to create an amorphous layer. At sufficiently high temperatures, an amorphous layer can no longer be formed.

Implantation damage can be removed by an "annealing" step. Following implantation, the wafer is heated to a temperature between 800 and 1000 °C for approximately 30 min. At this temperature, silicon atoms can move back into lattice sites, and impurity atoms can enter substitutional sites in the lattice. After the annealing cycle,

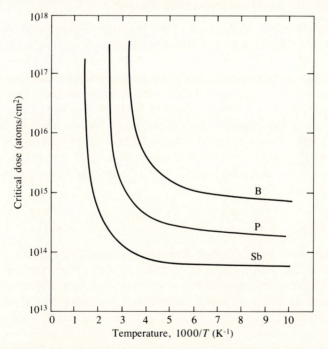

Fig. 5.9 A plot of the dose required to form an amorphous layer on silicon versus reciprocal target temperature. Arsenic falls between phosphorus and antimony. Copyright 1970 by Plenum Publishing Corporation. Reprinted with permission from ref. [6].

nearly all of the implanted dose becomes electrically active, except for impurity concentrations exceeding $10^{19}/cm^3$.

Annealing cycles for 30 min at temperatures approaching 1000 °C can cause considerable spreading of the implant by diffusion. It has been found that truly amorphous layers can actually be annealed at lower temperatures through the process of solid-phase epitaxy. The crystalline substrate seeds recrystallization of the amorphous layer, and epitaxial growth can proceed as rapidly as 500 Å/min at 600 °C. During solid-phase epitaxy, impurity atoms are incorporated into substitutional sites, and full activation is achieved at low temperatures.

Low-energy arsenic implantations produce shallow amorphous layers that can be annealed using solid-phase epitaxy to yield shallow, abrupt junctions that are ideal for VLSI structures. Boron, however, is so light that it does not produce an amorphous layer even at relatively high doses, unless the substrate is deliberately cooled (see Fig. 5.9). Today, boron is often implanted using ions of the heavier BF_2 molecule. The lower-velocity implant results in shallow layers that can be annealed under solid-phase-epitaxy conditions.

5.5.3 Deviations from the Gaussian Theory

If we take a detailed look at the shape of the implanted impurity distribution, we find deviations from the ideal Gaussian profile. When light ions, such as boron, impact atoms of the silicon target, they experience a relatively large amount of backward scattering and fill in the distribution on the surface side of the peak, as in Fig. 5.10. Heavy atoms, such as arsenic, experience a larger amount of forward scattering and tend to fill in the profile on the substrate side of the peak. A number of methods have been proposed for mathematical modeling of this behavior, such as the use of Pearson Type-IV distributions.[8] However, for common implant energies below 200 keV, the Gaussian theory provides a useful model of the impurity distribution. This is particularly true since the forward and backward scattering tend to alter the tails of the distribution where the concentration is well below the peak value.

5.6 SUMMARY

Ion implantation uses a high-voltage accelerator to introduce impurity atoms into the surface of the silicon wafer, and it offers many advantages over deposition by high-temperature diffusion. Ion implantation is a low-temperature process minimizing impurity movement by diffusion, which has become very important to VLSI fabrication. Low-temperature processing also permits the use of a wide variety of materials as barrier layers to mask the implantation. Photoresist, oxide, nitride, aluminum, and other metal films can all be used, adding important increased flexibility to process design.

Ion implantation also permits the use of a much wider range of impurity species than diffusion. In principle, any element that can be ionized can be introduced into the wafer using implantation. Implantation offers much tighter control of the dose introduced into

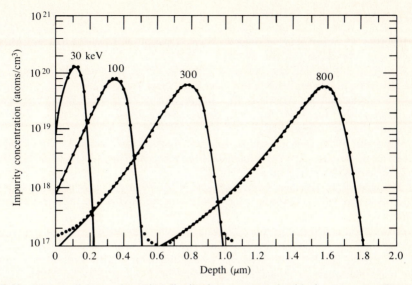

Fig. 5.10 Measured boron impurity distributions compared with four-moment (Pearson IV) distribution functions. The boron was implanted into amorphous silicon without annealing. Reprinted with permission from Philips Journal of Research.[8]

the wafer, and a much wider range of doses can be reproducibly achieved than possible with diffusion.

Diffused profiles almost always have the maximum impurity concentration at the surface. Ion-implantation techniques can be used to produce new profiles with subsurface peaks or retrograde profiles which decrease toward the wafer surface. Implantation can introduce impurities into very shallow layers near the surface, again a significant advantage for VLSI structures.

The main disadvantage of ion implantation is the cost of the equipment, which may exceed $2 million for a production machine. Also, the ion implanter has trouble achieving high doses ($>10^{16}/cm^3$) in a time reasonable for high-volume production. New high-current machines are being developed to overcome this latter problem. Overall, the flexibility and process control achievable with ion implantation have far outweighed the disadvantage of cost, and ion implantation is used routinely for state-of-the-art bipolar and MOS integrated-circuit fabrication.

Ion implantation results in profiles which can be modeled by a Gaussian distribution. The depth and width of the distribution depend on both the ion species and the energy of the implantation. In order to prevent channeling, implantation is normally performed at an angle of approximately 7° off the normal to the surface.

The implantation process damages the surface, and an annealing step is required to remove the effects of the damage. Low doses may result in the need for annealing at 800 to 1000 °C for 30 min. However, if the surface layer has become amorphous, annealing can be achieved through solid-phase epitaxy at temperatures of only 600 °C.

REFERENCES

[1] J. Lindhard, M. Scharff, and H. Schiott, "Range Concepts in Heavy Ion Ranges," Mat.-Fys. Med. Dan. Vid. Selsk, *33*, No. 14, 1963.

[2] J. F. Gibbons, W. S. Johnson, and S. W. Mylroie, *Projected Range in Semiconductors*, Second Edition, Dowden, Hutchinson, and Ross, New York, 1975.

[3] (a) L. Pauling and R. Hayward, *The Architecture of Molecules*, W. H. Freeman, San Francisco, 1964. (b) S. M. Sze, Ed., *Semiconductor Devices Physics and Technology*, McGraw-Hill, New York, 1985.

[4] S. Furukawa, H. Matsumura, and H. Ishiwara, "Lateral Distribution Theory of Implanted Ions," in S. Namba, Ed., *Ion Implantation in Semiconductors*, Japanese Society for the Promotion of Science, Kyoto, p. 73 (1972).

[5] G. Dearnaley, J. H. Freeman, G. A. Card, and M. A. Wilkins, "Implantation Profiles of ^{32}P Channeled into Silicon Crystals," Canadian Journal of Physics, *46*, 587–595 (March 15, 1968).

[6] F. F. Morehead and B. L. Crowder, "A Model for the Formation of Amorphous Si by Ion Implantation," p. 25–30, in Eisen and Chadderton (see Source Listing 4).

[7] B. L. Crowder and F. F. Morehead, Jr., "Annealing Characteristics of *n*-type Dopants in Ion-Implanted Silicon," Applied Physics Letters, *14*, 313–315 (May 15, 1969).

[8] W. K. Hofker, "Implantation of Boron in Silicon," Philips Research Reports Supplements, No. 8, 1975.

[9] J. F. Gibbons, "Ion-Implantation in Semiconductors — Part I: Range Distribution Theory and Experiment," Proceedings of the IEEE, *56*, 295–319 (March, 1968).

[10] J. F. Gibbons, "Ion Implantation in Semiconductors — Part II: Damage Production and Annealing," Proceedings of the IEEE, *60*, 1062–1096 (September, 1972).

[11] T. Hirao, G. Fuse, K. Inoue, S. Takayanagi, Y. Yaegashi, S. Ichikawa, and T. Izumi, "Electrical Properties of Si Implanted with As Through SiO_2 Films," Journal of Applied Physics, *51*, 262–268 (January, 1980).

SOURCE LISTING

(1) J. W. Mayer, L. Eriksson, and J. A. Davies, *Ion-Implantation in Semiconductors*, Academic Press, New York, 1970.

(2) G. Dearnaley, J. H. Freeman, R. S. Nelson, and J. Stephen, *Ion-Implantation*, North-Holland, New York, 1973.

(3) G. Carter and W. A. Grant, *Ion-Implantation of Semiconductors*, John Wiley & Sons, New York, 1976.

(4) F. Eisen and L. Chadderton, Eds., *Ion Implantation in Semiconductors*, First International Conference (Thousand Oaks, CA), Gordon and Breach, New York, 1970.

(5) I. Ruge and J. Graul, Eds., *Ion Implantation in Semiconductors*, Second International Conference (Garmisch-Partenkirchen, West Germany), Springer-Verlag, Berlin, 1972.

(6) B. L. Crowder, Ed., *Ion Implantation in Semiconductors*, Third International Conference (Yorktown Heights, NY), Plenum, New York, 1973.

(7) S. Namba, Ed., *Ion Implantation in Semiconductors*, Fourth International Conference (Osaka, Japan), Plenum, New York, 1975.

(8) F. Chernow, J. Borders, and D. Bruce, Eds., *Ion Implantation in Semiconductors*, Fifth International Conference (Boulder, CO), Plenum, New York, 1976.

PROBLEMS

5.1 Boron is implanted with an energy of 60 keV through a 0.25-μm layer of silicon dioxide. The implanted dose is $1 \times 10^{14}/\text{cm}^2$.

(a) Find the boron concentration at the silicon–silicon dioxide interface.

(b) Find the dose in silicon.

(c) Determine the junction depth if the background concentration is $3 \times 10^{15}/\text{cm}^3$.

5.2 An arsenic dose of $1 \times 10^{12}/\text{cm}^2$ is implanted through a 50-nm layer of silicon dioxide with the peak of the distribution at the Si-SiO$_2$ interface. A silicon nitride film on top of the silicon dioxide is to be used as a barrier material in the regions where arsenic is not desired. How thick should the nitride layer be if the background concentration is $1 \times 10^{15}/\text{cm}^3$?

5.3 An implantation will be used for the predeposition step for a boron base diffusion. The final layer is to be 5 μm deep with a sheet resistance of 125 ohms per square ($N_B = 10^{16}/\text{cm}^3$).

(a) What is the dose required from the ion implanter if the boron will be implanted through a thin silicon dioxide layer so that the peak of the implanted distribution is at the silicon–silicon dioxide interface?

(b) What drive-in time is required to produce the final base layer at a temperature of 1100 °C?

5.4 (a) Use Irvin's curves to find the sheet resistance of a boron layer implanted completely below the surface in n-type silicon ($N_B = 10^{15}/\text{cm}^3$). Assume the layer has a peak concentration of $1 \times 10^{19}/\text{cm}^3$, and the range and straggle are 1.0 μm and 0.11 μm, respectively. (Hint: Think about conductors in parallel.)

(b) What is the dose of this implantation?

(c) What was the energy used for this ion implantation?

(d) At what depths are pn junctions located?

5.5 The source and drain regions of a self-aligned n-channel polysilicon-gate MOS transistor are to be formed using arsenic implantation. The dimensions of a cross section of the device are given in Fig. P5.5. Calculate the channel shrinkage caused by lateral straggle if the peak concentration of the implantation is $10^{20}/\text{cm}^3$ and the substrate doping is $10^{16}/\text{cm}^3$. Assume that the channel region is in the silicon immediately below the oxide. Use $R_p = 0.1$ μm, $\Delta R_p = 0.04$ μm, and $\Delta R_\perp = 0.022$ μm.

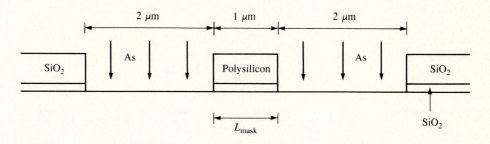

Fig. P5.5

5.6 An implanted profile is formed by two boron implantations. The first uses an energy of 100 keV and the second an energy of 200 keV. The peak concentration of each distribution is $5 \times 10^{18}/\text{cm}^3$. Draw a graph of the composite profile and find the junction depth(s) if the phosphorus background concentration is $10^{16}/\text{cm}^3$. What are the doses of the two implant steps.

5.7 A high energy (5 MeV) is used to implant oxygen deep below the silicon surface in order to form a buried SiO_2 layer. Assume that the desired SiO_2 layer is to be 0.2 μm wide.

(a) What is the oxygen dose required to be implanted in silicon?

(b) What beam current is required if a 125-mm-diameter wafer is to be implanted in 15 min?

(c) How much power is being supplied to the ion beam? Discuss what effects this implantation may have on the wafer.

5.8 The threshold voltage of a NMOS transistor may be increased by ion implantation of boron into the channel region. For shallow implantations, the voltage shift is given approximately by $\Delta V_T = qQ/C_{ox}$ where Q is the boron dose and $C_{ox} = \varepsilon_0/X_0$. X_0 is the oxide thickness and ε_0 is the permittivity of silicon dioxide: $3.9 \times (8.854 \times 10^{-14}$ F/cm). What boron dose is required to shift the threshold by 0.75 V if the oxide thickness is 40 nm?

5.9 An ion implanter has a beam current of 10 μA. How long does it take to implant a boron dose of $10^{15}/\text{cm}^2$ into a wafer with a diameter of 125 mm?

5.10 Write a computer program to calculate the sheet resistance of an arbitrary Gaussian layer in silicon.

6 / Film Deposition

Fabrication processes involve many steps in which thin films of various materials are deposited on the surface of the wafer. This chapter presents a survey of deposition processes, including evaporation, chemical vapor deposition, and sputtering, which are used to deposit metals, silicon and polysilicon, and dielectrics such as silicon dioxide and silicon nitride. Evaporation and sputtering require vacuum systems operating at low pressure, whereas chemical vapor deposition and epitaxy can be performed at either reduced or atmospheric pressure. An overview of vacuum systems and some results from the theory of ideal gases are also presented in this chapter.

6.1 EVAPORATION

Physical evaporation is one of the oldest methods of depositing metal films. Aluminum and gold are heated to the point of vaporization, and then evaporate to form a thin film covering the surface of the silicon wafer. In order to control the composition of the deposited material, evaporation is performed under vacuum conditions.

Figure 6.1 shows a basic vacuum deposition system consisting of a vacuum chamber, a mechanical roughing pump, a diffusion pump or turbomolecular pump, valves, vacuum gauges, and other instrumentation. In operation, the roughing valve is opened first, and the mechanical pump lowers the vacuum chamber pressure to an intermediate vacuum level of approximately 1 Pascal (Pa[†]). If a higher vacuum level is needed, the roughing valve is closed, and the foreline and high-vacuum valves are opened. The roughing pump now maintains a vacuum on the output of the diffusion pump. A liquid-nitrogen (77 K) cold trap is used with the diffusion pump to reduce the pressure in the vacuum chamber to approximately 10^{-4} Pa. Ion and thermocouple gauges are used to monitor the pressure at a number of points in the vacuum system, and several other valves are used as vents to return the system to atmospheric pressure.

[†]1 atm = 760 mm Hg = 760 torr = 1.013×10^5 Pa. 1 Pa = 1 N/m^2 = 0.0075 torr.

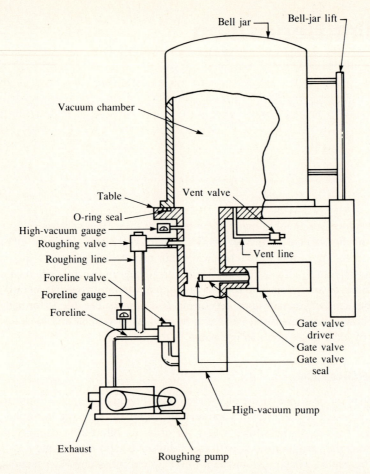

Fig. 6.1 Typical vacuum system used for evaporation including vacuum chamber, roughing pump, high-vacuum pump, and various valves and vacuum gauges. Copyright, 1987, McGraw-Hill Book Company; reprinted with permission from ref. [5].

6.1.1   Kinetic Gas Theory

Gases behave in an almost ideal manner at low pressure and are well described by the ideal gas law. Pressure P, volume V, and temperature T of one mole of a gas are related by

$$PV = N_{av}kT \tag{6.1}$$

in which $k^{\dagger}$ is Boltzmann's constant and N_{av} is Avogadro's number (6.02×10^{23}

$^{\dagger}k$ (Boltzmann's constant) $= 1.38 \times 10^{-23}$ J/K $= 1.37 \times 10^{-22}$ atm-cm^3/K.

molecules/mole). The concentration of gas molecules is given by

$$n = N_{av}/V = P/kT \tag{6.2}$$

In some systems, the surface of the substrate must be kept extremely clean prior to deposition. The presence of even a small amount of oxygen or other elements will result in formation of a contamination layer on the surface of the substrate. The rate of formation of this layer is determined from the impingement rate of gas molecules hitting the substrate surface and is related to the pressure by

$$\Phi = P/\sqrt{2\pi mkT} \text{ (molecules/cm}^2\text{-sec)} \tag{6.3}$$

where m is the mass of the molecule. This can be reduced to

$$\Phi = 2.63 \times 10^{20} \, P/\sqrt{MT} \text{ (molecules/cm}^2\text{-sec)} \tag{6.4}$$

where P is the pressure in Pa and M is the molecular weight (e.g., $M = 32$ for oxygen molecules). If we assume that each molecule sticks as it contacts the surface, then the time required to form a monolayer on the surface is given by

$$t = N_s/\Phi = N_s\sqrt{2\pi mkT}/P \tag{6.5}$$

where N_s is the number of molecules/cm^2 in the layer.

Example: Suppose the residual pressure of oxygen in the vacuum system is 1 Pa. How long does it take to deposit one atomic layer of oxygen on the surface of the wafer at 300 K?

Solution: The radius of an oxygen molecule is approximately 3.6 Å. If we assume close packing of the molecules on the surface, there will be approximately 2.2×10^{14} molecules/cm^2. At 300 K and 1 Pa, the impingement rate for oxygen is 2.7×10^{18} molecules/cm^2-sec. One monolayer is deposited in 82 μsec.

Pressure and temperature also determine another important film-deposition parameter called the *mean free path*, λ. The mean free path of a gas molecule is the average distance the molecule travels before it collides with another molecule. λ is given by

$$\lambda = kT/\sqrt{2}\,\pi P d^2 \tag{6.6}$$

in which d is the diameter of the gas molecule and is in the range of 2 to 5 Å. Evaporation is usually done at a background pressure near 10^{-4} Pa. At this pressure, a 4-Å molecule has a mean free path of approximately 60 m. Thus, during aluminum evaporation, for example, aluminum molecules do not interact with the background gases and tend to travel in a straight line from the evaporation source to the deposition target.

On the other hand, sputtering, which will be discussed in Section 6.4, uses argon gas at a pressure of approximately 100 Pa. Using the same radius results in a mean free path of only 60 μm. Thus the material being deposited tends to scatter often with the argon atoms and arrives at the target from random directions.

6.1.2 Filament Evaporation

The simplest evaporator consists of a vacuum system containing a filament which can be heated to high temperature. In Fig. 6.2a, small loops of a metal such as aluminum are hung from a filament formed of a refractory (high-temperature) metal such as tungsten. Evaporation is accomplished by gradually increasing the temperature of the filament until the aluminum melts and wets the filament. Filament temperature is then raised to evaporate the aluminum from the filament. The wafers are mounted near the filament and are covered by a thin film of the evaporating material.

Although filament evaporation systems are easy to set up, contamination levels can be high, particularly from the filament material. In addition, evaporation of composite materials cannot be easily controlled using a filament evaporator. The material with the lowest melting point tends to evaporate first, and the deposited film will not have the

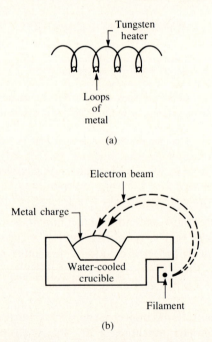

Fig. 6.2 Two forms of evaporation sources. (a) Filament evaporation, in which loops of wire hang from a heated filament; (b) electron-beam source in which a beam of electrons is focused on a metal charge. The beam is bent in a magnetic field.

same composition as the source material. Thick films are difficult to achieve since a limited supply of material is contained in the metal loops.

6.1.3   Electron-Beam Evaporation

In electron-beam (E-beam) evaporation systems (see Fig. 6.2b), the high-temperature filament is replaced with an electron beam. A high-intensity beam of electrons, with an energy up to 15 keV, is focused on a source target containing the material to be evaporated. The energy from the electron beam melts a region of the target. Material evaporates from the source and covers the silicon wafers with a thin layer.

The growth rate using a small planar source is given by

$$G = \frac{m}{\pi \rho r^2} \cos \phi \, \cos \theta \, (\text{cm/sec}) \tag{6.7}$$

for the geometrical setup in Fig. 6.3. ϕ is the angle measured from the normal to the plane of the source, and θ is the angle of the substrate relative to the vapor stream. ρ and m are the density (g/cm^3) and mass evaporation rate (g/sec), respectively, of the material being deposited.

For batch deposition, a planetary substrate holder (Fig. 6.4) consisting of rotating sections of a sphere is used. Each substrate is positioned tangential to the surface of the sphere with radius r_o, as in Fig. 6.3. Application of some geometry yields

$$\cos \theta = \cos \phi = r/2r_o \tag{6.8}$$

For the planetary substrate holder, G becomes independent of substrate position:

$$G = m/4\pi \rho r_o^2 \tag{6.9}$$

The wafers are mounted above the source and are typically rotated around the source during deposition to ensure uniform coverage. The wafers are also often radiantly heated to improve adhesion and uniformity of the evaporated material. The source material sits in a water-cooled crucible, and its surface only comes in contact with the electron beam during the evaporation process. Purity is controlled by the purity of the original source material. The relatively large size of the source provides a virtually unlimited supply of material for evaporation, and the deposition rate is easily controlled by changing the current and energy of the electron beam.

One method of monitoring the deposition rate uses a quartz crystal which is covered by the evaporating material during deposition. The resonant frequency of the crystal shifts in proportion to the thickness of the deposited film. By monitoring the resonant frequency of the crystal, the deposition rate may be measured with an accuracy of better than 1 Å/sec. Dual electron beams with dual targets may be used to coevaporate composite materials in E-beam evaporation systems.

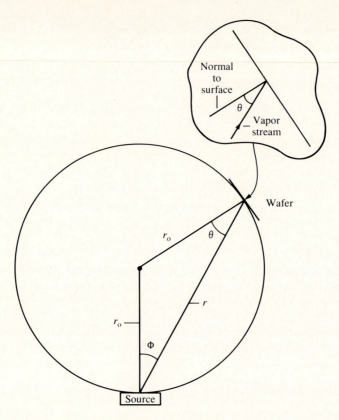

Fig. 6.3   Geometry for evaporation in a system using a planetary substrate holder.

X-ray radiation can be generated in an electron-beam system for acceleration voltages exceeding 5 to 10 keV. Substrates may suffer some radiation damage from both energetic electrons and X-rays. The damage can usually be annealed out during subsequent process steps. However, the radiation effects are of great concern to MOS process designers, and so sputtering has replaced electron-beam evaporation in many steps in manufacturing processes.

6.1.4   Flash Evaporation

Flash evaporation uses a fine wire as the source material, and a high-temperature ceramic bar is used to evaporate the wire. The wire is fed continuously and evaporates on contact with the ceramic bar. Flash evaporation can produce relatively thick films, as in an E-beam system, without problems associated with radiation damage.

Fig. 6.4   Photograph of an E-beam evaporation system with a planetary substrate holder which rotates simultaneously around two axes.

6.1.5 Shadowing and Step Coverage

Because of the large mean free paths of gas molecules at low pressure, evaporation techniques tend to be directional in nature, and shadowing of patterns and poor step coverage can occur during deposition. Figure 6.5 illustrates the shadowing phenomenon which can occur with closely spaced features on the surface of an integrated circuit. In the fully shadowed region, there will be little deposition. In the partially shadowed region, there will be variation in film thickness. To minimize these effects, the planetary substrate holder of the electron-beam system continuously rotates the wafers during the film deposition.

6.2 SPUTTERING

Sputtering is achieved by bombarding a target with energetic ions, typically Ar^+. Atoms at the surface of the target are knocked loose and transported to the substrate, where deposition occurs. Electrically conductive materials such as Al, W, and Ti can use a dc power source, in which the target acts as the cathode in a diode system. Sputtering of dielectrics such as silicon dioxide or aluminum oxide requires an RF power source to supply energy to the argon atoms. A diagram of a sputtering system is shown in Fig. 6.6.

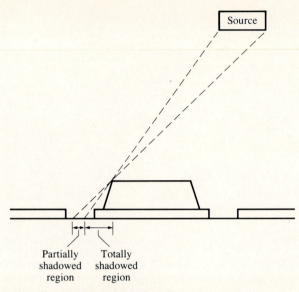

Fig. 6.5   An example of the shadowing problem that can occur in low-pressure vacuum deposition in which the molecular mean free path is large.

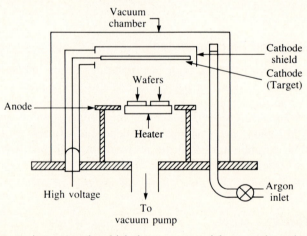

Fig. 6.6   A dc sputtering system in which the target material acts as the cathode of a diode and the wafers are mounted on the system anode.

In sputter deposition, there is a threshold energy which must be exceeded before sputtering occurs. The sputtering yield (Fig. 6.7) is the ratio of the number of atoms liberated from the target by each incident atom, and it increases rapidly with energy of the incident atoms. Systems are usually operated with an energy large enough to ensure a sputtering yield of at least unity.

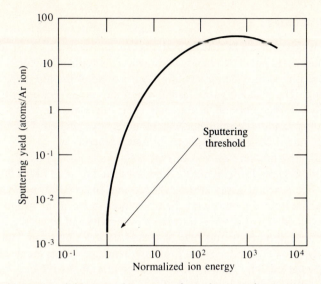

Fig. 6.7 Sputtering yield versus ion energy for a dc sputtering system using argon.

Sputtering can be used to deposit a broad range of materials. In addition, alloys may be deposited in which the film has the same composition as the target. An example is the Al-Cu-Si alloy commonly used for metallization in integrated circuits. (We will discuss this alloy in Chapter 7.) As one might expect, sputtering results in the incorporation of some argon into the film, and heating of the substrate up to 350 °C can occur during the deposition process. Sputtering can also give excellent coverage of the sharp topologies often encountered in integrated circuits.

Sputter etching (a reversal of the sputter deposition process) can be used to clean the substrate prior to film deposition, and the sputter etching process is often used to clean contact windows prior to metal deposition. Etching removes any residual oxide from the window and improves the contact between the metal and the underlying material.

6.3 CHEMICAL VAPOR DEPOSITION

Chemical vapor deposition (CVD) forms thin films on the surface of a substrate by thermal decomposition and/or reaction of gaseous compounds. The desired material is deposited directly from the gas phase onto the surface of the substrate. Polysilicon, silicon dioxide, and silicon nitride are routinely deposited using CVD techniques. In addition, refractory metals such as tungsten (W) can also be deposited using CVD.

Chemical vapor deposition can be performed at pressures for which the mean free path for gas molecules is quite small, and the use of relatively high temperatures can result in excellent conformal step coverage over a broad range of topological profiles.

6.3.1 CVD Reactors

Several different types of CVD reactor systems are shown in Fig. 6.8. In Fig. 6.8a, a continuous atmospheric-pressure reactor is shown. This type of reactor is often used for deposition of the silicon dioxide passivation layer as one of the last steps in integrated-circuit processing. The reactant gases flow through the center section of the reactor and are contained by nitrogen curtains at the ends. The substrates can be fed continuously through the system, and large-diameter wafers are easily handled. However, high gas-flow rates are required by the atmospheric-pressure reactor.

The hot-wall, low-pressure system of Fig. 6.8b is commonly used to deposit poly-silicon, silicon dioxide, and silicon nitride, and is referred to as an LPCVD (low-pressure CVD) system. The reactant gases are introduced into one end of a three-zone furnace tube and are pumped out the other end. Temperatures range from 300 to 1150 °C, and the pressure is typically 30 to 250 Pa. Excellent uniformity can be obtained with LPCVD systems, and several hundred wafers may be processed in a single run. Hot-wall systems have the disadvantage that the deposited film simultaneously coats the inside of the tube. The tube must be periodically cleaned or replaced to minimize problems with particulate matter. In spite of this problem, hot-wall LPCVD systems are in widespread use throughout the semiconductor industry.

CVD reactions can also take place in a plasma reactor, as shown in Fig. 6.8c. Formation of the plasma permits the reaction to take place at low temperatures, which is a primary advantage of plasma-enhanced CVD (PECVD) processes. In the parallel-plate system, the wafers lie on a grounded aluminum plate which serves as the bottom electrode for establishing the plasma. The wafers can be heated up to 400 °C using high-intensity lamps or resistance heaters. The top electrode is a second aluminum plate placed in close proximity to the wafer surface. Gases are introduced along the outside of the system, flow radially across the wafers, and are pumped through an exhaust in the center. An RF signal is applied to the top plate to establish the plasma. The capacity of this type of system is limited, and wafers must be loaded manually. A major problem in VLSI fabrication is particulate matter that may fall from the upper plate onto the wafers.

The furnace-plasma system in Fig. 6.8d can handle a large number of wafers at one time. A special electrode assembly holds the wafers parallel to the gas flow. The plasma is established between alternating groups of electrodes supporting the wafers.

6.3.2 Polysilicon Deposition

Silicon is deposited in an LPCVD system using thermal decomposition of silane:

$$SiH_4 \xrightarrow{600\ °C} Si + 2H_2 \qquad (6.10)$$

Low-pressure systems (25 to 150 Pa) use either 100% silane or 20 to 30% silane diluted with nitrogen. A temperature between 600 and 650 °C results in deposition of polysilicon material at a rate of 100 to 200 Å/min. A less commonly used deposition occurs between

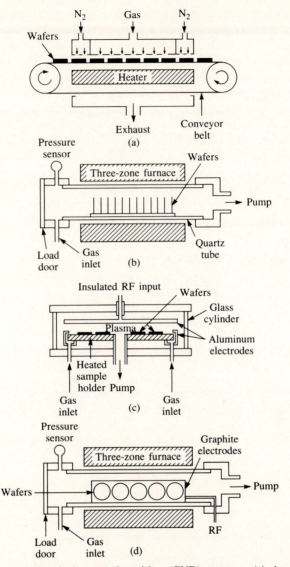

Fig. 6.8 Four types of chemical vapor deposition (CVD) systems. (a) Atmospheric-pressure reactor; (b) hot-wall LPCVD system using a three-zone furnace tube; (c) parallel-plate plasma-enhanced CVD system; (d) PECVD system using a three-zone furnace tube. Copyright, 1983, Bell Telephone Laboratories, Inc. Reprinted by permission from ref. [2].

850 and 1050 °C in a hydrogen atmosphere. The higher temperature overcomes a reduction in deposition rate caused by the hydrogen carrier gas.

Polysilicon can be doped by diffusion or ion implantation or during deposition (in situ) by the addition of dopant gases such as phosphine, arsine, or diborane. The addition of diborane greatly increases the deposition rate, whereas the addition of phosphine or arsine substantially reduces the deposition rate.

Polysilicon is often deposited as undoped material and is then doped by diffusion. High-temperature diffusion occurs much more rapidly in polysilicon than in single-crystal silicon, and the polysilicon film is typically saturated with the dopant to achieve as low a resistivity as possible for interconnection purposes. Resistivities of 0.01 to 0.001 ohm-cm can be achieved in diffusion-doped polysilicon. Ion implantation typically yields a lower active-impurity density in the polysilicon film, and ion-implanted polysilicon exhibits a resistivity about ten times higher than that achieved by high-temperature diffusion.

6.3.3 Silicon Dioxide Deposition

Silicon dioxide films can be deposited using a variety of reactions and temperature ranges, and the films can be doped or undoped. Phosphorus-doped oxide can be used as a passivation layer over a completed integrated circuit or as the insulating medium in multilevel metal processes (which will be discussed in the next chapter). Silicon dioxide containing 6 to 8% phosphorus by weight will soften and flow at temperatures between 1000 and 1100 °C. This "P-glass reflow" process is often used to improve step coverage and provide a smoother topography for later process steps. SiO_2 with lower concentrations of phosphorus will not reflow properly, and higher concentrations can corrode aluminum if moisture is present. Oxide doped with 5 to 15% by weight of various dopants can also be used as a diffusion source.

Deposition of silicon dioxide over aluminum must occur at a temperature below the silicon-aluminum eutectic point of 577 °C (see Chapter 7). A reaction between silane and oxygen is commonly used between 300 and 500 °C:

$$SiH_4 + O_2 \longrightarrow SiO_2 + 2H_2 \tag{6.11}$$

The oxide may be doped with phosphorus using phosphine:

$$4PH_3 + 5O_2 \longrightarrow 2P_2O_5 + 6H_2 \tag{6.12}$$

Oxide passivation layers can be deposited at atmospheric pressure using the continuous reactor of Fig. 6.8a, or they can be deposited at reduced pressure in an LPCVD system, as in Fig. 6.8b.

Deposition of SiO_2 films prior to metallization can be performed at higher temperatures, which gives a wider choice of reactions and results in better uniformity and step coverage. For example, a dichlorosilane reaction with nitrous oxide in an LPCVD system at approximately 900 °C,

$$SiCl_2H_2 + 2N_2O \longrightarrow SiO_2 + 2N_2 + 2HCl \tag{6.13}$$

is used to deposit insulating layers of SiO_2 on wafer surfaces.

Decomposition of the vapor produced from a liquid source, tetraethylorthosilicate (TEOS), can also be used in an LPCVD system between 650 and 750 °C.

$$Si(OC_2H_5)_4 \longrightarrow SiO_2 + \text{by-products} \tag{6.14}$$

Deposition based on the decomposition of TEOS provides excellent uniformity and step coverage. Oxide doping may be accomplished in the LPCVD systems by adding phosphine, arsine, or diborane.

A comparison of some of the properties of various CVD oxides is given in Table 6.1.

6.3.4 Silicon Nitride Deposition

As discussed in Chapter 3, silicon nitride is used as an oxidation mask in recessed oxide processes. Silicon nitride is also used as a final passivation layer because it provides an excellent barrier to both moisture and sodium contamination. Composite films of oxide and nitride are being investigated for use as very thin gate insulators in scaled VLSI devices, and they are also used as the gate dielectric in electrically programmable memory devices.

Both silane and dichlorosilane will react with ammonia to produce silicon nitride. The silane reaction occurs between 700 and 900 °C at atmospheric pressure:

$$3SiH_4 + 4NH_3 \longrightarrow Si_3N_4 + 12H_2 \tag{6.15}$$

Dichlorosilane is used in an LPCVD system between 700 and 800 °C:

$$3SiCl_2H_2 + 4NH_3 \longrightarrow Si_3N_4 + 6HCl + 6H_2 \tag{6.16}$$

Table 6.1 Properties of Various Deposited Oxides. (After ref. [2].)

Source	Deposition Temperature (°C)	Composition	Conformal Step Coverage	Dielectric Strength (MV/cm)	Etch Rate (Å/min) [100:1 H₂O:HF]
Silane	450	$SiO_2(H)$	No	8	60
Dichlorosilane	900	$SiO_2(Cl)$	Yes	10	30
TEOS	700	SiO_2	Yes	10	30
Plasma	200	$SiO_{1.9}(H)$	No	5	400

Thermal growth of silicon nitride is also possible but not very practical. Silicon nitride will form when silicon is exposed to ammonia at temperatures between 1000 and 1100 °C, but the growth rate is very low.

Plasma systems may also be used for the deposition of silicon nitride. Silane will react with a nitrogen discharge to form plasma nitride (SiN):

$$2SiH_4 + N_2 \longrightarrow 2SiNH + 3H_2 \qquad (6.17)$$

Silane will also react with ammonia in an argon plasma:

$$SiH_4 + NH_3 \longrightarrow SiNH + 3H_2 \qquad (6.18)$$

LPCVD films are hydrogen-rich, containing up to 8% hydrogen. Plasma deposition does not produce stoichiometric silicon nitride films. Instead, the films contain as much as 20 to 25% hydrogen. LPCVD films have high internal tensile stresses, and films thicker than 2000 Å may crack because of this stress. On the other hand, plasma-deposited films have much lower tensile stresses.

The resistivity (10^{16} ohm-cm) and dielectric strength (10 MV/cm) of the LPCVD nitride film are better than those of most plasma films. Resistivity of plasma nitride can range from 10^6 to 10^{15} ohm-cm, depending on the amount of nitrogen in the film, while the dielectric strength ranges between 1 and 5 MV/cm.

6.3.5 CVD Metal Deposition

Many metals can be deposited by CVD processes. Molybdenum (Mo), tantalum (Ta), titanium (Ti), and tungsten (W) are all of interest in today's processes because of their low resistivity and their ability to form silicides with silicon (see Chapter 7). Aluminum can be deposited from a metallorganic compound such as tri-isobutyl aluminum, but this technique has not been commonly used because many other excellent methods of aluminum deposition are available.

Tungsten can be deposited by thermal, plasma, or optically assisted decomposition of WF_6:

$$WF_6 \longrightarrow W + 3F_2 \qquad (6.19)$$

or through reduction with hydrogen:

$$WF_6 + 3H_2 \longrightarrow W + 6HF \qquad (6.20)$$

Mo, Ta, and Ti can be deposited in an LPCVD system through reaction with hydrogen. The reaction is the same for all three metals:

$$2MCl_5 + 5H_2 \longrightarrow 2M + 10HCl \qquad (6.21)$$

where M stands for any one of the three metals mentioned above.

6.4 EPITAXY

Chemical vapor deposition processes can be used to deposit silicon onto the surface of a silicon wafer. Under appropriate conditions, the silicon wafer acts as a seed crystal, and a single-crystal silicon layer is grown on the surface of the wafer. The growth of a crystalline silicon layer from the vapor phase is called *vapor-phase epitaxy* (VPE), and it is the most common form of epitaxy used in silicon processing. In addition, *liquid-phase epitaxy* (LPE) and *molecular-beam epitaxy* (MBE) are being used widely in GaAs technology.

Epitaxial growth was first used in integrated-circuit processing to grow single-crystal *n*-type layers on *p*-type substrates for use in standard buried-collector bipolar processing. More recently, it has been introduced into CMOS VLSI processes where lightly doped layers are grown on heavily doped substrates of the same type (*n* on n^+ or *p* on p^+) to help suppress a circuit-failure mode called *latchup*.

6.4.1 Vapor-Phase Epitaxy

Silicon epitaxial layers are commonly grown with silicon deposited from the gas phase. A basic model for the process is given in Fig. 6.9. At the silicon surface, the flux J_s of gas molecules is determined by

$$J_s = k_s N_s \tag{6.22}$$

in which k_s is the surface-reaction rate constant and N_s is the surface concentration of the molecule involved in the reaction. In the steady state, this flux must equal the flux J_g of molecules diffusing in from the gas stream. The flux J_g may be approximated by

$$J_g = (\overline{D}_g/\delta)(N_g - N_s) = h_g(N_g - N_s) \tag{6.23}$$

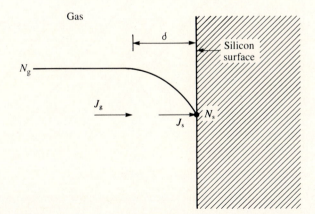

Fig. 6.9 Schematic model for the epitaxial growth process.

in which $\overline{D}_g$ is an effective diffusion constant for the gas molecule and δ is the distance over which the diffusion is taking place. The ratio $\overline{D}_g/\delta$ is called the *vapor-phase mass-transfer coefficient*, h_g. Equating J_s and J_g yields the flux impinging on the surface of the wafer. The growth rate is equal to the flux divided by the number N of molecules incorporated per unit volume of film.

$$v = \frac{J_s}{N} = \frac{k_s h_g}{k_s + h_g}\frac{N_g}{N} \qquad (6.24)$$

If $k_s \gg h_g$, then growth is said to be mass-transfer-limited, and

$$v = h_g\frac{N_g}{N} \qquad (6.25)$$

If $h_g \gg k_s$, then growth is said to be surface-reaction-limited, and

$$v = k_s\frac{N_g}{N} \qquad (6.26)$$

Figure 6.10 shows epitaxial growth rate as a function of temperature. Chemical reactions at the surface tend to follow an Arrhenius relationship characterized by an activation

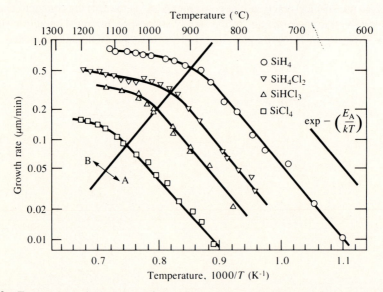

Fig. 6.10 Temperature dependence of the silicon epitaxial growth process for four different sources. The growth rate is surface-reaction-limited in region A and is mass-transfer-limited in region B. Reprinted with permission from Philips Journal of Research from ref. [3].

energy E_A, whereas the mass-transfer process tends to be independent of temperature. These two regions show up clearly in this figure. At low temperatures, the growth rate follows an Arrhenius relationship with an activation energy of approximately 1.5 eV. At higher temperatures, the growth rate becomes independent of temperature. In order to have good growth-rate control and to minimize sensitivity to variations in tempera-ture, epitaxial growth conditions are usually chosen to yield a mass-transfer-limited growth rate.

Three common types of VPE reactors, the horizontal, vertical, and barrel systems, are shown in Fig. 6.11. The susceptor that supports the wafers is made of graphite and is heated by RF induction in the horizontal and vertical reators and by radiant heating in the barrel reactor.

Silicon tetrachloride ($SiCl_4$), silane (SiH_4), dichlorosilane (SiH_2Cl_2), and trichloro-silane ($SiHCl_3$) have all been used for silicon VPE. Silicon tetrachloride has been widely used in industrial processing:

$$SiCl_4(gas) + 2H_2(gas) \longleftrightarrow Si(solid) + 4HCl(gas) \qquad (6.27)$$

This reaction takes place at approximately 1200 °C and is reversible. If the carrier gas coming into the reactor contains hydrochloric acid, etching of the surface of the silicon wafer can occur. This in situ etching process can be used to clean the wafer prior to the start of epitaxial deposition.

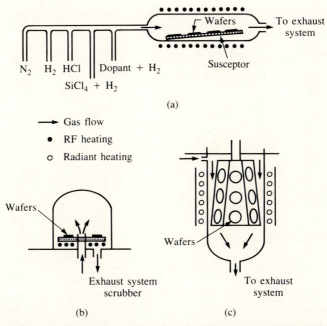

Fig. 6.11   (a) Horizontal, (b) pancake, and (c) barrel susceptors commonly used for vapor-phase epitaxy. Copyright, 1985, John Wiley & Sons, Inc., with permission from ref. [1].

A second reaction competes with the epitaxial deposition process:

$$SiCl_4(gas) + Si(solid) \longleftrightarrow 2SiCl_2(gas) \tag{6.28}$$

This second reaction also etches the silicon from the wafer surface. If the concentration of $SiCl_4$ is too high, etching of the wafer surface will take place rather than epitaxial deposition. Figure 6.12 shows the effect of $SiCl_4$ concentration on the growth of epitaxial silicon. The growth rate initially increases with increasing $SiCl_4$ concentration, peaks, then decreases. Eventually, growth stops and the etching process becomes dominant. If the growth rate is too high, a polysilicon layer is deposited rather than a layer of single-crystal silicon.

Epitaxial growth can also be achieved by the pyrolytic decomposition of silane:

$$SiH_4 \xrightarrow{650 \,°C} Si + 2H_2 \tag{6.29}$$

The reaction is not reversible and takes place at low temperatures. In addition, it avoids the formation of HCl gas as a reaction by-product. However, careful control of the reactor is needed to prevent formation of polysilicon rather than single-crystal silicon layers. The presence of any oxidizing species in the reactor can also lead to contamination of the epitaxial layer by silica dust.

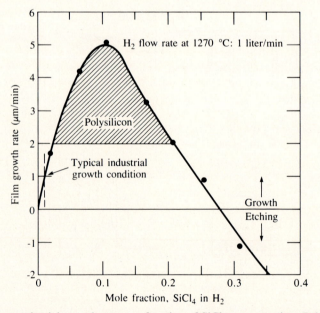

Fig. 6.12   Silicon epitaxial growth rate as a function of $SiCl_4$ concentration. Polysilicon deposition occurs for growth rates exceeding 2 μm/min. Etching of the surface will occur for mole fraction concentrations exceeding 28%. Copyright, 1985, John Wiley & Sons, Inc., with permission from ref. [1].

6.4.2 Doping of Epitaxial Layers

Epitaxial layers may be doped during the growth process by adding impurities to the gas used for deposition. Arsine, diborane, and phosphine are the most convenient sources of the common impurities. The resistivity of the epitaxial layer is controlled by varying the partial pressure of the dopant species in the gas supplied to the reactor. The addition of arsine or phosphine tends to slow down the rate of epitaxial growth, while the addition of diborane tends to enhance the epitaxial growth rate.

Lightly doped epitaxial layers are often grown on more heavily doped substrates, and "autodoping" of the epitaxial layer can occur during growth. Impurities can evaporate from the wafer or may be liberated by chlorine etching of the surface during deposition. The impurities are incorporated into the gas stream, resulting in doping of the growing layer. As the epitaxial layer grows, less dopant is released from the wafer into the gas stream, and the impurity profile eventually reaches a constant level determined by the doping in the gas stream.

During deposition, the substrate also acts as a source of impurities which diffuse into the epitaxial layer. This "out-diffusion" will be discussed more fully in the next section. Both autodoping and out-diffusion cause the transition from the doping level of the substrate to that of the epitaxial layer to be less abrupt than desired. The effects of autodoping and out-diffusion are illustrated in Fig. 6.13.

6.4.3 Buried Layers

Out-diffusion is a common problem that occurs with the buried layer in bipolar transistors. In order to reduce the resistance in series with the collector of the bipolar transistor, heavily doped n-type regions are diffused into the substrate prior to the growth of an n-type epitaxial layer. During epitaxy, impurities diffuse upward from the heavily doped buried-layer regions.

Diffusion of impurities from the substrate during epitaxial growth is modeled by the diffusion equation with a moving boundary,[4] as in Fig. 6.14:

$$D \frac{\partial^2 N}{\partial x^2} = \frac{\partial N}{\partial t} + v_x \frac{\partial N}{\partial x} \tag{6.30}$$

in which v_x is the rate of growth of the epitaxial layer.

Two specific solutions of eq. (6.30) are applicable to epitaxial layer growth. The first case is the growth of an undoped epitaxial layer on a uniformly doped substrate. The boundary conditions are $N(x, 0) = N_s = N(\infty, t)$, and the flux $J_x = (h + v_x)N(0, t)$ where h is the mass-transfer coefficient which characterizes the escape rate of dopant atoms from the silicon into the gas. Normally, $h \ll v_x$. A change of variables from x to $x' = x - v_x t$ simplifies eq. (6.30) and gives an approximate solution for $N(x, t)$:

$$N_1(x, t) = \frac{N_s}{2} \left[1 + \text{erf} \ \frac{x - x_{\text{epi}}}{2\sqrt{D_s t}} \right] \tag{6.31}$$

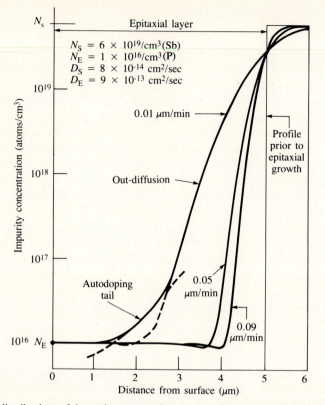

Fig. 6.13 Redistribution of impurity atoms due to gas-phase autodoping and impurity out-diffusion during epitaxial layer growth. Out-diffusion is calculated using eq. (6.33) for epitaxial growth of a phosphorus-doped layer at 1150 °C over an antimony-doped buried layer with a surface concentration of 6×10^{19}/cm^3. The three curves are for growth rates of 0.01, 0.05, and 0.09 μm/min. For clarity, the effects of autodoping are shown on only one curve.

Eq. (6.31) assumes that the epitaxial layer growth rate greatly exceeds the rate of movement of the diffusion front. Eq. (6.31) is the exact solution for diffusion from one semi-infinite layer into a second semi-infinite layer.

The second case is the growth of a doped epitaxial layer on an undoped substrate. The boundary conditions for this case are $N(0, t) = N_E$, and $N(\infty, t) = 0 = N(x, 0)$. The solution of eq. (6.30) for these boundary conditions is:

$$N_2(x, t) = \frac{N_E}{2} \left[\text{erfc} \frac{x - x_{\text{epi}}}{2\sqrt{D_E t}} + \exp \frac{v_x x}{D_E} \text{erfc} \frac{x + x_{\text{epi}}}{2\sqrt{D_E t}} \right] \qquad (6.32)$$

in which $x_{\text{epi}} = v_x t$ is the epitaxial layer thickness. Superposition of the solutions for

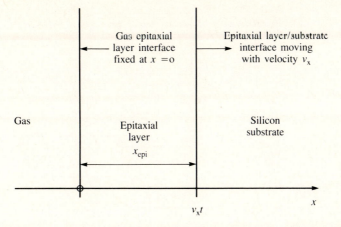

Fig. 6.14 Geometrical structure of moving boundary value problem which models the epitaxial growth process.

these two cases gives a good approximation to diffusion which occurs during epitaxial growth:

$$N(x, t) = N_1(x, t) + N_2(x, t) \qquad (6.33)$$

N_S represents the doping in the substrate, and N_E is the doping intentionally introduced into the epitaxial layer. D_S and D_E represent the diffusion coefficients of the impurity species in the substrate and epitaxial layer, respectively. Figure 6.13 shows diffusion profiles for a phosphorus-doped epitaxial layer grown at various rates on an antimony-doped substrate. The curves were produced using eq. (6.33).

An additional problem occurs during epitaxial growth. The oxidation and lithographic processing steps used during formation of a buried layer result in a step of as much as 0.2 μm around the perimeter of the buried layer. Epitaxial growth on this nonplanar surface causes the pattern to shift during growth, as illustrated in Fig. 6.15. Pattern shift is difficult to predict, may be as large as the epitaxial layer thickness, and must be accounted for during the design of subsequent mask levels.

6.4.4 Liquid-Phase and Molecular-Beam Epitaxy

In liquid-phase epitaxy, the substrate is brought into contact with a solution containing the material to be deposited in liquid form. The substrate acts as a seed for material crystallizing directly from the solute. Growth rates typically range between 0.1 and 1 μm/min.

In the molecular-beam epitaxy process, the crystalline layer is formed by deposition from a thermal beam of atoms or molecules. Deposition is performed in ultrahigh-

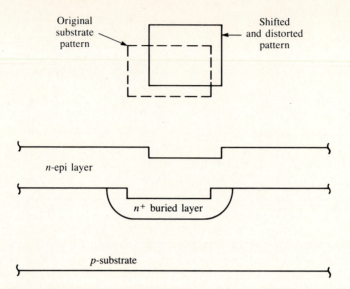

Fig. 6.15 Pattern shift during epitaxial growth over an n^+ buried layer. The original pattern is shifted and distorted in shape.

vacuum conditions (10^{-8} Pa). Substrate temperatures during MBE range from 400 to 900 °C, and the growth rate is relatively low (0.001 to 0.3 μm/min). The epitaxial layer is grown atomic layer by atomic layer, and many unique device structures can be fabricated by changing the material which is deposited between one layer and the next.

The throughput of MBE is relatively low at the present time. Plasma-assisted CVD processes, which promise to give many of the benefits of MBE with much higher throughput, are presently being investigated in research laboratories.

6.5 SUMMARY

Thin films of a very broad range of materials are used in integrated-circuit fabrication. This chapter has presented an overview of film-deposition techniques including physical evaporation, chemical vapor deposition (CVD), epitaxial growth, and sputtering. Most of these processes are performed at low pressure, and this chapter has presented an introduction to vacuum systems and a review of some important aspects of ideal gas theory.

Physical evaporation using filament or electron-beam evaporators can be used to deposit metals and other materials which can easily be melted. E-beam systems can operate at high power levels and melt high-temperature metals. However, E-beam evaporation may result in radiation damage to thin oxide layers at the surface of the wafer. In addition, it is difficult to deposit material compounds and alloys using evaporation. Finally, gas molecules at low pressures have large mean free paths, and evaporation has problems with shadowing and poor step coverage during film deposition.

Sputtering uses energetic ions such as argon to bombard a target material and dislodge atoms from the surface of the target. The dislodged atoms are deposited on the surface of the wafer. Direct-current sputtering systems can be used to deposit conductive materials, and RF sputtering can be used to deposit insulators. Sputtering can be used to deposit composite materials in which the deposited film maintains the same composition as the source material. Sputtering also uses higher pressures than evaporation. The much shorter mean free paths which result yield a deposition with freedom from shadowing and much better step coverage.

Low-pressure and atmospheric chemical vapor deposition (CVD) systems deposit films from chemical reactions taking place in a gas stream passing over the wafer. Polysilicon, silicon dioxide, silicon nitride, and metals can all be deposited using CVD techniques. A special type of CVD deposition called epitaxy results in the growth of single-crystal silicon films on the surface of silicon wafers. Out-diffusion and autodoping cause problems with impurity profile control during epitaxial layer growth.

In a modern bipolar or MOS fabrication process, one can expect to find evaporation, sputtering, and chemical vapor deposition techniques all used somewhere in the process flow.

REFERENCES

[1] S. M. Sze, *Semiconductor Devices — Physics and Technology,* John Wiley & Sons, New York, 1985.

[2] A. C. Adams, "Dielectric and Polysilicon Film Deposition," Chapter 3 in S. M. Sze, Ed., *VLSI Technology,* McGraw-Hill, New York, 1983.

[3] F. C. Eversteyn, "Chemical-Reaction Engineering in Semiconductor Industry," *Philips Research Reports, 29,* 45–66 (February, 1974).

[4] A. B. Glaser and G. E. Subak-Sharpe, *Integrated Circuit Engineering,* p. 205–209, Addison-Wesley, Reading, MA, 1979.

[5] W. S. Ruska, *Microelectronic Processing,* McGraw-Hill, New York, 1987.

FURTHER READING

1. J. L. Vossen and W. Kern, Eds., *Thin Film Processes,* Academic Press, New York, 1978.

2. J. F. O'Hanlon, *A User's Guide to Vacuum Technology,* John Wiley & Sons, New York, 1980.

3. L. Holland, *Vacuum Deposition of Thin Films,* John Wiley & Sons, New York, 1961.

4. A. S. Grove, *Physics and Technology of Semiconductor Devices,* John Wiley & Sons, New York, 1967.

5. H. C. Theuerer, "Epitaxial Silicon Films by Hydrogen Reduction of SiCl$_4$," Journal of the Electrochemical Society, *108,* 649–653 (July, 1961).

6. C. O. Thomas, D. Kahng, and R. C. Manz, "Impurity Distribution in Epitaxial Silicon Films," Journal of the Electrochemical Society, *109,* 1055–1061 (November, 1962).

7. A. S. Grove, A. Roder, and C. T. Sah, "Impurity Distribution in Epitaxial Growth," Journal of Applied Physics, *36*, 802–810 (March, 1965).

8. D. Kahng, C. O. Thomas, and R. C. Manz, "Epitaxial Silicon Junctions," Journal of the Electrochemical Society, *110*, 394–400 (May, 1963).

9. W. H. Shepherd, "Vapor Phase Deposition and Etching of Silicon," Journal of the Electrochemical Society, *112*, 988–994 (October, 1965).

10. G. R. Srinivasan, "Autodoping Effects in Silicon Epitaxy," Journal of the Electrochemical Society, *127*, 1334–1342 (June, 1980).

11. J. C. Bean, "Silicon Molecular Beam Epitaxy as a VLSI Processing Technique," 1981 IEEE IEDM Proceedings, p. 6–13.

PROBLEMS

6.1 A silicon wafer sits on a bench in the laboratory at a temperature of 300 K and a pressure of 1 atm. Assume that the air consists of 100% oxygen. How long does it take to deposit one atomic layer of oxygen on the wafer surface, assuming 100% adhesion?

6.2 Calculate the impingement rate and mean free path for oxygen molecules ($M = 32$) at 300 K and a pressure of 10^{-4} Pa. What is this pressure in torr?

6.3 An ultrahigh vacuum system operates at a pressure of 10^{-8} Pa. What is the concentration of residual air molecules in the chamber at 300 K?

6.4 The partial pressure of a material being deposited in a vacuum system must be well above the residual background gas pressure if reasonable deposition rates are to be achieved. What must the partial pressure of aluminum be to achieve a deposition rate of 100 nm/min? Assume close packing of spheres with a diameter of 5 Å, 100% adhesion of the impinging aluminum and 300K.

6.5 A wafer 100 mm in diameter is mounted in an electron-beam evaporation system in which the spherical radius is 40 cm. Use eq. (6.7) to estimate the worst-case variation in film thickness between the center and edges of the wafer for an evaporated aluminum film 1 μm thick.

6.6 A MBE system must operate under ultrahigh vacuum conditions to prevent the formation of undesired atomic layers on the surface of the substrate. What pressure of oxygen can be permitted at 300K if formation of a monolayer of contamination can be permitted after the sample has been in the chamber for no less than 4 hr?

6.7 (a) Calculate the growth rate of a silicon layer from a $SiCl_4$ source at 1200 °C. Use $h_g = 1$ cm/sec, $k_s = 2 \times 10^6 \exp(-1.9/kT)$ cm/sec, and $N_g = 3 \times 10^{16}$ atoms/cm^3. (For silicon, $N = 5 \times 10^{22}$/cm^3.)

(b) What is the change in growth rate if the temperature is increased by 25 °C?

(c) At what temperature does $k_s = h_g$? What is the growth rate at this temperature?

(d) What is the value of E_A in Fig. 6.10?

6.8 Use eqs. (6.31) and (6.32) to model the case of a 10-μm n-type epitaxial layer ($N_E = 1 \times 10^{16}$/cm^3) grown on a p-type substrate ($N_S = 1 \times 10^{18}$/cm^3). Plot the impurity profile in the epitaxial layer and substrate assuming that the layer was grown at a rate of 0.2 μm/min at a temperature of 1200 °C. Assume boron and phosphorus are the impurities. Find the location of the *pn* junction.

6.9 Compare and discuss the advantages and disadvantages of evaporation, sputtering, and chemical vapor deposition.

6.10 A 1-kg source of aluminum is used in an E-beam evaporation system. How many 100-mm wafers can be coated with a 1-μm Al film before the source material is exhausted? Assume that 15% of the evaporated aluminum actually coats a wafer. (The rest is deposited on the inside of the electron-beam system.)

6.11 A silicon wafer 75 mm in diameter is centered 100 mm above a small planar evaporation source. Calculate the ratio of thickness between the center and edges of the wafer using eq. (6.7), following a 1-μm film deposition.

7 / Interconnections and Contacts

The previous six chapters have focused on the various processes used to fabricate semiconductor devices in the silicon substrate. To complete the formation of an integrated circuit, one must interconnect the devices and finally get connections to the world outside the silicon chip. Until the 1970s, integrated circuits had two possible levels of interconnection: diffusions and metallization. The use of polysilicon as a gate material in MOS devices added a third level useful for interconnecting devices and circuits.

In this chapter we discuss the various forms of interconnections and the problems associated with making good contacts between metal and silicon. Refractory metal silicides and multilevel metallization used in VLSI processes are discussed, and an additional method for depositing patterned films, called *liftoff,* is also introduced.

7.1 INTERCONNECTIONS IN INTEGRATED CIRCUITS

As we found in previous chapters, aluminum, polysilicon, and diffused regions are all easily isolated from each other using an insulating layer of silicon dioxide. Thus today's ICs have three different materials that may cross over each other. To be useful as an interconnect, the materials must also provide as low a sheet resistance as possible in order to minimize voltage drops along the interconnect lines as well as to minimize propagation delay caused by the resistance and capacitance of the line. Finally, low-resistance "ohmic" contacts must be made between the materials, and the interconnection lines must be reliable throughout long-term operation.

Figure 7.1 shows a simple MOS logic circuit which illustrates how polysilicon, aluminum, and diffused interconnections may cross over and/or contact each other. Aluminum is used to make contact to diffusions and polysilicon, and diffusions in various regions have been extended and merged together to form interconnections.

In this technology, polysilicon lines and diffused lines can only be connected together using the metal level. Improved circuit density can often be achieved by using "butted contacts" between polysilicon and diffusions or by changing the process to introduce "buried contacts" directly between the polysilicon and diffused layers. These two techniques will be examined later in this chapter.

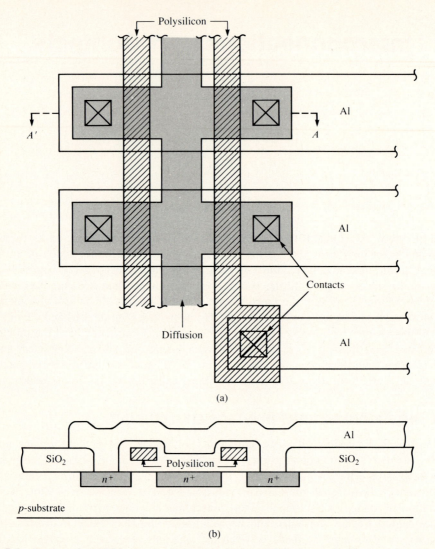

Fig. 7.1 Portion of a MOS logic circuit showing the use of diffusion, polysilicon, and aluminum interconnections. (a) Top view; (b) cross section through A'-A.

7.2 METAL INTERCONNECTIONS AND CONTACT TECHNOLOGY

The requirement for low-resistivity materials leads one to immediately consider metals for use as interconnections, and both aluminum and gold have been used with silicon IC processing. Gold requires the use of a multilayer metal sandwich involving other metals such as titanium or tungsten. In addition, gold can be troublesome because it is a very rapid diffuser and produces deep-level recombination centers in silicon. Aluminum is compatible with silicon IC processing and is the most common material in use today. It is relatively inexpensive, adheres well to silicon dioxide, and has a bulk resistivity of 2.7 μohm-cm. However, care must be exercised to avoid a number of problems associated with the formation of good aluminum contacts to silicon.

7.2.1 Ohmic Contact Formation

We desire to form "ohmic" contacts between the metal and semiconductor. True ohmic contacts would exhibit a straight-line *I-V* characteristic with a low value of resistance (Fig. 7.2a), as opposed to the *I-V* characteristic of a rectifying contact shown in Fig. 7.2b. Figure 7.2c shows an *I-V* characteristic which is more representative of a practical ohmic contact to silicon. Although nonlinear near the origin, it develops only a small voltage across the contact at normal current levels.

Figure 7.3 shows a number of ways in which aluminum may contact semiconductor regions during device fabrication. Aluminum contact to *p*-type silicon normally results in a good ohmic contact for doping levels exceeding $10^{16}/cm^3$. However, a problem arises in trying to contact *n*-type silicon, as shown in Fig. 7.3b. For lightly doped *n*-type material, aluminum can form a metal-semiconductor "Schottky-barrier" diode rather than an ohmic contact. In order to prevent this rectifying contact from forming, an n^+ diffusion is placed between the aluminum and any lightly doped *n*-type regions, as in Fig. 7-3c. The resulting contact has an *I-V* characteristic similar to that in Fig. 7.2c. This technique was used in the bipolar process shown in Fig. 1.6.

7.2.2 Aluminum-Silicon Eutectic Behavior

Silicon melts at a temperature of 1412 °C, and pure aluminum melts at 660 °C. However, aluminum and silicon together exhibit "eutectic" characteristics in which mixture of the

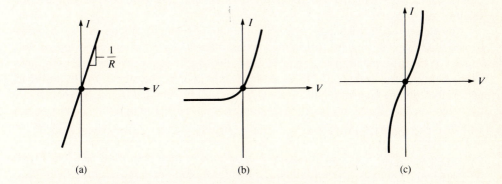

Fig. 7.2 *I-V* characteristics of contacts between integrated-circuit materials. (a) Ideal ohmic contact; (b) rectifying contact; (c) practical nonlinear "ohmic" contact.

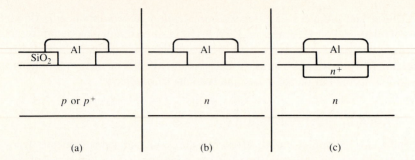

Fig. 7.3   Three possible types of aluminum contacts to silicon. (a) Aluminum to *p*-type silicon forms an ohmic contact with an *I-V* characteristic approximating that in Fig. 7.2a; (b) aluminum to *n*-type silicon can form a rectifying contact (Schottky barrier diode) like that in Fig. 7.2b; (c) aluminum to n^+ silicon yields a contact similar to that in Fig. 7.2c.

two materials lowers the melting point of the composite material to below that of either element. Figure 7.4 shows the phase diagram of the aluminum-silicon system at a pressure of 1 atm. The minimum melting temperature, the "eutectic temperature," is 577 °C and corresponds to a 88.7% Al, 11.3% Si composition. Because of the relatively low eutectic temperature of the Al-Si system, aluminum must be introduced into the IC process sequence after all high-temperature processing has been completed.

7.2.3   Aluminum Spiking and Junction Penetration

In order to ensure good contact formation, aluminum is normally annealed in an inert atmosphere at a temperature of 450 to 500 °C following deposition and patterning. Although this temperature is well below the eutectic temperature for silicon and aluminum, silicon still diffuses into the aluminum. The diffusion leads to a major problem associated with the formation of aluminum contacts to silicon, particularly for shallow junctions.

Anywhere a contact is made between aluminum and silicon, silicon will be absorbed by the aluminum during the annealing process. The amount of silicon absorbed will depend on the time and temperature involved in the annealing process, as well as the area of the contact (see Problem 7.4). To make matters worse, the silicon is not absorbed uniformly from the contact region. Instead, it tends to be supplied from a few points. As the silicon is dissolved, spikes of aluminum form and penetrate into the silicon contact region. If the contact is to a shallow junction, the spike may cause a junction short, as in Fig. 7.5.

The inset in Fig. 7.4 gives the solubility of silicon in aluminum. Between 400 °C and the eutectic temperature, the solubility of silicon in aluminum ranges from 0.25 to 1.5% by weight. To solve the spiking problem, silicon may be added to the aluminum film during deposition by coevaporation from two targets, or sputter deposition can be used with an aluminum target which contains approximately 1% silicon. Both of these

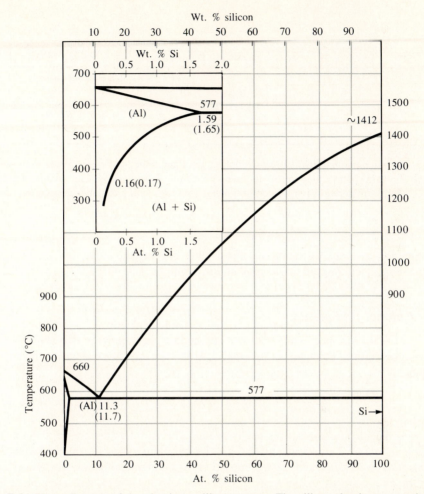

Fig. 7.4 Phase diagram of the aluminum-silicon system. The silicon-aluminum eutectic point occurs at a temperature of 577 °C. At contact-alloying temperatures between 450 and 500 °C, aluminum will absorb from 0.5 to 1% silicon. Copyright, 1958, McGraw-Hill Book Company, with permission from ref. [1].

techniques deposit a layer in which the aluminum demand for silicon is satisfied, and the metallization does not absorb silicon from the substrate during subsequent annealing steps.

Another way to prevent spiking is to place a barrier material between the aluminum and silicon, as shown in Fig. 7.5. One possibility is to deposit a thin layer of polysilicon prior to aluminum deposition. The polysilicon will then supply the silicon needed to saturate the aluminum. Another alternative is to use a metal as a barrier. The metal must form a low-resistance contact with silicon, not react with aluminum, and be compatible

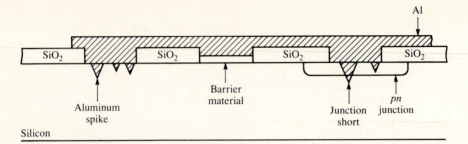

Fig. 7.5 Aluminum spiking which occurs during aluminum-silicon alloying. Aluminum spikes can cause shorts in shallow junctions. Aluminum containing 1% silicon is often used to eliminate spiking. A barrier material of polysilicon or a metal such as titanium can also be used to prevent spiking.

with other process steps. A number of metals have been used by various semiconductor manufacturers, including platinum, palladium, titanium, and tungsten.

7.2.4 Contact Resistance

There is a small resistance associated with an ohmic contact between two materials. To a first approximation, the "contact resistance" R_c is inversely proportional to the area of the contact:

$$R_c = \rho_c/A \qquad (7.1)$$

where ρ_c is the specific contact resistivity in ohm-cm^2 and A is the area of the contact. For example, a 2×2 μm contact with $\rho_c = 1$ μohm-cm^2 yields a contact resistance of 25 ohms. Figure 7.6 shows the contact resistivity as a function of annealing temperature for several aluminum-silicon systems. It is evident why the 450 °C annealing process is used following aluminum deposition. Also note that the use of polysilicon under aluminum to prevent junction spiking yields a much poorer value of ρ_c.

7.2.5 Electromigration

Metal interconnections in integrated circuits are operated at relatively high current densities, and a very interesting failure mechanism develops in aluminum and other conductors. *Electromigration* is the movement of atoms in a metal film due to momentum transfer from the electrons carrying the current. Under high-current-density conditions, metal-atom movement causes voids in some regions and metal pileup or "hillocks" in other regions, as shown in Fig. 7.7. Voids can eventually result in open circuits, and pileup can cause short circuits between closely spaced conductors.

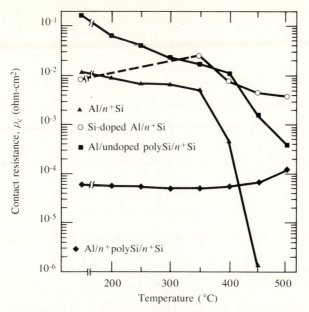

Fig. 7.6 Contact resistivity of a variety of aluminum-silicon systems. An alloying temperature of 450 °C is typically used to obtain low-contact resistance for Al-Si contacts. Reprinted with permission from *Solid-State Electronics,* Vol. 23, p. 255–262, M. Finetti et al., "Aluminum-Silicon Ohmic Contact on Shallow n$^+$/p Junctions," Copyright 1980, Pergamon Press, Ltd.

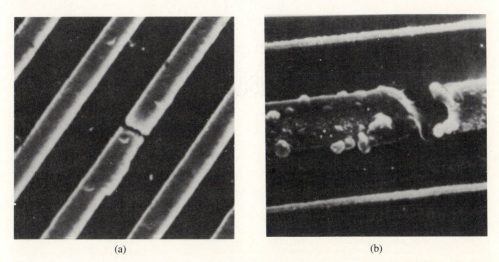

Fig. 7.7 Scanning electron micrographs of aluminum interconnection failure caused by electro-migration. (a) Sputtered aluminum with 0.5% copper; (b) evaporated aluminum with 0.5% copper. Copyright, 1980, IEEE. Reprinted with permission from ref. [3].

The mean time to failure (MTF) of a conductor due to electromigration has been experimentally related to current density, J, and temperature by

$$\text{MTF} \propto (J^{-2}) \exp(E_A/kT) \tag{7.2}$$

where E_A is an activation energy with a typical value of 0.4 to 0.5 eV for aluminum.

The most common method of improving aluminum resistance to electromigration is to add a small percentage of a heavier metal such as copper. Targets composed of 95% Al, 4% Cu, and 1% Si are routinely used in sputter deposition systems. The aluminum-copper-silicon alloy films simultaneously provide electromigration resistance and eliminate aluminum spiking.

7.3 DIFFUSED INTERCONNECTIONS

Diffused conductors with low sheet resistances represent the second available interconnect medium in basic integrated-circuit technology. From Fig. 4.16 we can see that the minimum resistivity is approximately 1000 μohm-cm. For shallow structures measuring about 1 μm, the minimum obtainable sheet resistance is typically between 10 and 20 ohms per square. Such sheet resistances are obviously much higher than that of metal, and one must be selective in the use of diffusions for signal or power distribution.

The diffused line must really be modeled as a distributed RC structure, illustrated in Fig. 7.8, when signal propagation is considered. The resistance, R, of diffused regions was discussed in detail in Chapter 4, and C represents capacitance of the reverse-biased

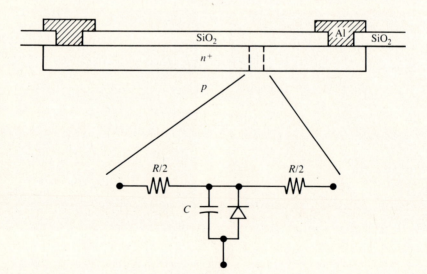

Fig. 7.8 A lumped circuit model for a small section of an n^+ diffusion. The *RC* line delay limits the use of diffusions for high-speed signal distribution.

pn junction formed between the diffused region and the substrate. Heavily doped diffusions are normally used for interconnection purposes and can be approximated by a
one-sided step junction in which the depletion layer extends predominantly into the
substrate. The capacitance per unit area is given by

$$C = \sqrt{\frac{qN_S K_s \varepsilon_o}{2(\phi_{bi} + V_R)}} \qquad \phi_{bi} = (kT/q) \ln\left(\frac{N_S}{n_i}\right) + 0.56 \text{ V} \qquad (7.3)$$

where N_S is the substrate doping, ϕ_{bi} is the built-in potential of the junction, and V_R is
the reverse bias applied to the junction.

The relatively large *RC* product of long diffused lines results in substantial time delay
for signals propagating down such a line. Hence, diffusions are more useful in interconnecting adjacent devices in integrated circuits. Figure 7.9 shows a three-input NMOS
NOR-gate in which the source diffusions of the three input transistors are merged
together as one diffusion. The three drains of the input devices, as well as the source of
the depletion-mode load device, are also merged together as one diffusion. Figure 7.1
shows an example of the use of long diffused interconnection regions in a programmable-
logic-array (PLA) structure.

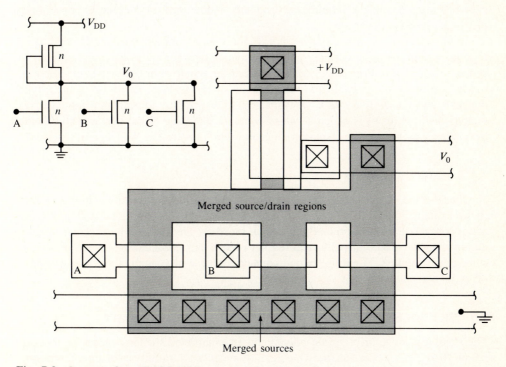

Fig. 7.9 Layout of an NMOS NOR-gate showing device interconnection through merging of
adjacent source and drain diffusions.

7.4   POLYSILICON INTERCONNECTIONS AND BURIED CONTACTS

Heavily doped n-type polysilicon is the primary MOS transistor gate material in use today, and it provides an additional layer of interconnection which is easily insulated from other layers by thermal oxidation or insulator deposition. This extra level of interconnection greatly facilitates the layout of compact digital integrated circuits. Thin, heavily doped polysilicon layers have a minimum resistivity of approximately 300 μohm-cm, and they suffer from the same sheet-resistance problems associated with shallow diffused interconnections (typically 20 to 30 ohms per square). Polysilicon lines have substantial capacitance to the substrate and exhibit RC delay problems similar to those of diffused interconnections.

7.4.1   Buried Contacts

In the polysilicon-gate processes presented thus far, the polysilicon acts as a barrier material during ion implantation or diffusion. Thus, a diffusion can never pass beneath a polysilicon line. In addition, contact windows to the diffusions are not opened until after polysilicon deposition. It is therefore necessary to use a metal link to connect between polysilicon and diffusion, as in Fig. 7.10a. Interconnecting the diffusion to polysilicon in this manner requires two contact windows and an intervening space, both of which are wasteful of area.

In memory arrays, where density is extremely important, an extra mask step can be introduced into the process to permit direct contact between polysilicon and silicon, as shown in Fig. 7.10b. Prior to polysilicon deposition, windows are opened in the thin gate oxide, permitting the polysilicon to contact the silicon surface. Diffusion of the n-type dopant from the heavily doped n^+ polysilicon merges with the adjacent ion-implanted n^+ regions, and the result is called a *buried contact*. The edge of the contact exhibits the lowest resistance since the impurity concentration is greatest in that region.

7.4.2   Butted Contacts

Another method of conserving area is to form a "butted" contact as shown in Fig. 7.10c. In this example, polysilicon is aligned with the edge of the diffusion contact window, and metal connects the diffusion and polysilicon together. The butted contact saves area by eliminating the space normally required between separate contact windows. However, some manufacturers do not use butted contacts because of concern about the reliability of such contacts.

7.5   SILICIDES AND MULTILAYER-CONTACT TECHNOLOGY

The sheet resistance of both thin polysilicon and shallow diffusions cannot be reduced below 10 to 20 ohms per square, which greatly reduces their utility as an interconnection medium. Interconnect delays are beginning to limit the speed of VLSI circuits, and as

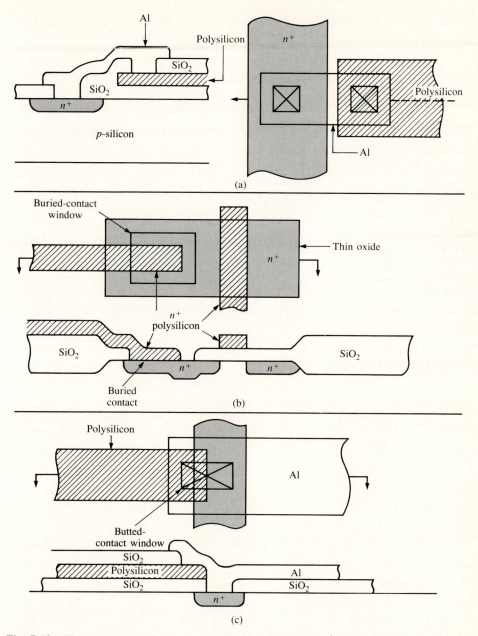

Fig. 7.10 Three techniques for interconnecting polysilicon and n^+ diffusion. (a) Normal aluminum link requiring two contact regions and an intervening space; (b) buried-contact structure; (c) butted-contact structure.

dice get larger and feature sizes get smaller, methods for improving these interconnections have had to be found.

7.5.1 Silicides, Polycides, and Salicides

A wide range of noble and refractory metals form compounds with silicon called *silicides,* and the sheet resistance of polysilicon and diffusion can be reduced by forming a low-resistivity, shunting silicide layer on their surfaces. A list of properties of possible silicides is given in Table 7.1. Several of the elements, including titanium, tungsten, platinum, and palladium, have been used in the formation of Schottky-barrier diodes in bipolar processes since the 1960s and are now used to form silicides for interconnection purposes.

A structure with a silicide formed on top of the polysilicon gate, often called a *polycide,* is shown in Fig. 7.11. A layer of the desired metal is deposited using evaporation, sputtering, or CVD techniques. Upon heating of the structure to a temperature between 600 and 1000 °C, the metal reacts with the polysilicon to form the desired silicide. Coevaporation, cosputtering, or sputtering of a composite target may be used to simultaneously deposit both silicon and metal onto the polysilicon surface prior to the thermal treatment or "sintering" step. Silicides have resistivities in the range of 15 to 50 μohm-cm.

Table 7.1 Properties of Some Silicides of Interest. Reprinted with permission of the American Institute of Physics from ref. [4].

Silicide	Starting Form	Sintering Temperature (°C)	Lowest Binary Eutectic Temperature (°C)	Specific Resistivity (μohm-cm)
$CoSi_2$	Metal on polysilicon	900	1195	18–25
	Cosputtered alloy	900		
$HfSi_2$	Metal on polysilicon	900	1300	45–50
$MoSi_2$	Cosputtered alloy	1000	1410	100
$NiSi_2$	Metal on polysilicon	900	966	50
	Cosputtered alloy	900		50–60
Pd_2Si	Metal on polysilicon	400	720	30–50
PtSi	Metal on polysilicon	600–800	830	28–35
$TaSi_2$	Metal on polysilicon	1000	1385	35–45
	Cosputtered alloy	1000		50–55
$TiSi_2$	Metal on polysilicon	900	1330	13–16
	Cosputtered alloy	900		25
WSi_2	Cosputtered alloy	1000	1440	70
$ZrSi_2$	Metal on polysilicon	900	1355	35–40

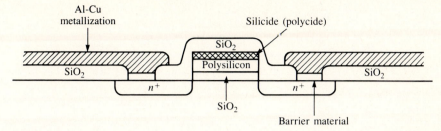

Fig. 7.11 MOS structure showing the use of a "polycide" to reduce the sheet resistance of the polysilicon gate material and a barrier material to prevent aluminum spiking through shallow source/drain junctions.

Another feature of silicide layers is the ability to oxidize the surface following silicide formation. At high temperature, silicon diffuses readily through the silicide layer and will combine with oxygen at the silicide surface to form an SiO_2 insulating layer.

The eutectic temperature of the silicide and silicon will limit the temperature of further processing steps, as in the case of aluminum. However, many silicides are stable at temperatures exceeding 1000 °C. Exceptions include the silicides of nickel (900 °C), platinum (800 °C), and palladium (700 °C).

Silicides are also used to reduce the effective sheet resistance of diffused interconnections. Figure 7.12 outlines a process for simultaneous formation of silicides on both the gate and source/drain regions of an MOS transistor. An oxide spacer is used to prevent silicide formation on the side of the gate, because such formation could cause a short between the gate and diffusions. The spacer is formed by first coating the surface with a CVD oxide, followed by a reactive-ion etching step. The oxide along the edge of the gate is thicker than over other regions, and some oxide is left on the side of the gate at the point when the oxide is completely removed from the source and drain regions and the top of the gate. Next, metal is deposited over the wafer. During sintering, silicide forms only in the regions where metal touches silicon or polysilicon. Unreacted metal may be removed with a selective etch which does not attack the silicide. The result is a silicide that is automatically self-aligned to the gate and source/drain regions. *Self-aligned silicides* are often called *salicides*.

7.5.2 Barrier Metals and Multilayer Contacts

Aluminum contacts to silicides suffer from the same pitting and spiking problems associated with direct contact to silicon. To circumvent these problems, an intermediate layer of metal is used that prevents silicon diffusion. Figure 7.13 shows the application of titanium-tungsten, TiW, as a barrier metal over the silicides in the contact regions of both bipolar and MOS technologies. The final contact consists of a sandwich of a silicide over the diffusion, followed by the TiW diffusion barrier, and completed with aluminum-copper interconnection metallization. Multilayer contact structures are becoming quite common in high-density, high-performance MOS and bipolar technologies.

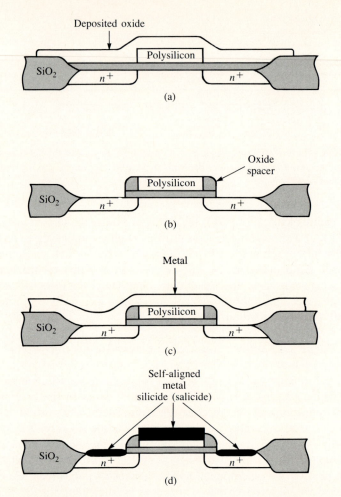

Fig. 7.12 Use of a self-aligned silicide ("salicide") in the formation of a MOS device. (a) Oxide is deposited over the normal MOS structure following polysilicon definition; (b) structure after reactive-ion etching leaving a sidewall oxide spacer; (c) metal is deposited over the structure and heated to form silicides; (d) unreacted metal is readily etched away, leaving silicide automatically aligned to gate and source/drain regions.

7.6 THE LIFTOFF PROCESS

The pattern definition processes which have been discussed previously have been "subtractive" processes, as illustrated in Fig. 7.14a. The wafer is completely covered with a thin film layer, which is selectively protected with a masking layer such as photoresist. Wet or dry etching then removes the thin film material from the unprotected areas.

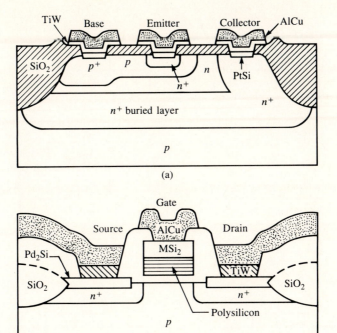

Fig. 7.13   Device cross sections showing the use of silicide contacts in (a) bipolar and (b) MOS devices. Reprinted with permission from Semiconductor International magazine, August 1985.[5] Copyright 1985 by Cahners Publishing Co., Des Plaines, IL.

The additive or *liftoff* process shown in Fig. 7.14b can also be used, in which the substrate is first covered with a photoresist layer patterned with openings where the final material is to appear. The thin film layer is deposited over the surface of the wafer. Any material deposited on top of the photoresist layer will be removed with the resist, leaving the patterned material on the substrate. For liftoff to work properly, there must be a very thin region or a gap between the upper and lower films. Otherwise tearing and incomplete liftoff will occur.

The masking patterns for the liftoff and subtractive processes are the negatives of each other. This can be achieved by changing the mask from dark field to light field or by changing from negative to positive photoresist.

7.7   MULTILEVEL METALLIZATION

A single level of metal simply does not provide sufficient capability to fully interconnect complex VLSI chips. Many processes now use two or three levels of polysilicon as well

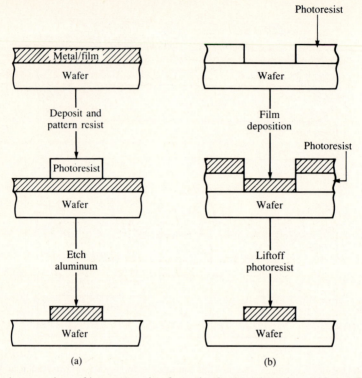

Fig. 7.14 A comparison of interconnection formation by (a) subtractive etching and (b) additive metal liftoff.

as several levels of metallization in order to ensure wirability and provide adequate power distribution.

A multilevel metal system is shown in Fig. 7.15. Standard processing is used through the deposition and patterning of the first level of metal. An interlevel dielectric, consisting of CVD or sputtered SiO_2, or a plastic-like material called *polyimide,* is then deposited over the first metal layer. The dielectric layer must provide good step coverage and should help smooth the topology. In addition, the layer must be free of pinholes and be a good insulator.

Next, vias are opened in the dielectric layer, and the second level of metallization is deposited and patterned. In the process in Fig. 7.15b, a via filling technique, which improves the overall topology, is used prior to deposition of each metal level. The dielectric deposition and metallization processes are repeated until the desired number of levels of interconnection are achieved. Integrated circuits with up to four levels of metal have been successfully fabricated using similar processes.

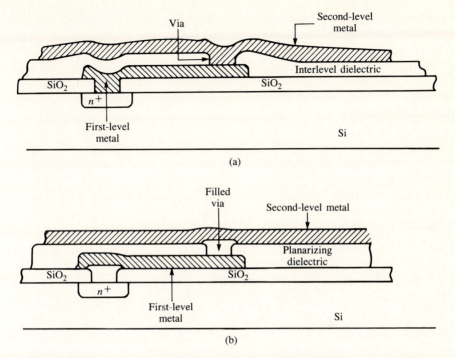

Fig. 7.15 (a) Basic two-level metallization process may use polyimide, oxide, or nitride as an interlevel dielectric; (b) additional process steps may be added to fill the vias with metal prior to each metal deposition in order to achieve a more planar structure.

7.8 SUMMARY

In this chapter we have explored the various types of interconnections used in modern integrated circuits, including diffusion, polysilicon, and metal. Diffusion and polysilicon have a relatively high sheet resistance, which often restricts their use to local interconnections. The formation of metal silicides on the surface of polysilicon lines and diffusions can substantially reduce the sheet resistance of these interconnections.

Problems relating to the formation of good ohmic contacts between aluminum and silicon have also been discussed. An n^+ layer is required between aluminum and n-type silicon to prevent formation of a Schottky-barrier diode instead of an ohmic contact. Aluminum penetration into silicon is a serious problem in forming contacts to shallow junctions. Metals such as tungsten and titanium are often used as silicon diffusion barriers to prevent aluminum penetration into contacts to silicon or silicides.

At high current densities, a failure mechanism called *electromigration* can cause open and short circuits to form in the metallization layers. Aluminum containing approximately 1% silicon and 4% copper is used to minimize aluminum spiking and electromigration, respectively.

Multilevel metal processes have been developed for integrated circuits which require more than one level of metallization. Some of today's processes contain up to three levels of polysilicon, and others use four levels of metallization.

REFERENCES

[1] M. Hansen and A. Anderko, *Constitution of Binary Alloys,* McGraw-Hill, New York, 1958.

[2] M. Finetti, P. Ostoja, S. Solmi, and G. Soncini, "Aluminum-Silicon Ohmic Contact on 'Shallow' n^+/p Junctions," Solid-State Electronics, *23,* 255–262 (March, 1980).

[3] S. Vaidya, D. B. Fraser, and A. K. Sinha, "Electromigration Resistance of Fine Line Al for VLSI Applications," Proceedings of 18th IEEE Reliability Physics Symposium, p. 165–167, 1980.

[4] S. P. Murarka, "Refractory Silicides for Integrated Circuits," Journal of Vacuum Science and Technology, *17,* 775–792 (July/August, 1980).

[5] P. S. Ho, "VLSI Interconnect Metallization–Part 3," Semiconductor International, 128–133 (August, 1985).

FURTHER READING

1. P. B. Ghate, J. C. Blair, and C. R. Fuller, "Metallization in Microelectronics," Thin Solid Films, *45,* 69–84 (August, 15, 1977).

2. G. L. Schnable and R. S. Keen, "Aluminum Metallization — Advantages and Limitations for Integrated Circuit Applications," Proceedings of the IEEE, *57,* 1570–1580 (September, 1969).

3. J. Black, "Physics of Electromigration," Proceedings of the 12th IEEE Reliability Physics Symposium, p. 142, 1974.

4. C. Y. Ting, "Silicide for Contacts and Interconnects," IEEE IEDM Digest, p. 110–113 (December, 1984).

5. S. P. Murarka, "Recent Advances in Silicide Technology," Solid-State Technology, *28,* 181–185 (September, 1985).

6. S. Sachdev and R. Castellano, "CVD Tungsten and Tungsten Silicide for VLSI Applications," Semiconductor International, p. 306–310 (May, 1985).

7. P. B. Ghate, J. C. Blair, C. R. Fuller, and G. E. McGuire, "Application of Ti:W Barrier Metallization for Integrated Circuits," Thin Solid Films, *53,* 117–128 (September 1, 1978).

8. R. A. M. Wolters and A. J. M. Nellissen, "Properties of Reactive Sputtered TiW," Solid-State Technology, *29,* 131–136 (February, 1986).

9. T. Sakurai and T. Serikawa, "Lift-off Metallization of Sputtered Al Alloy Films," Journal of the Electrochemical Society, *126,* 1257–1260 (July, 1979).

10. T. Batchelder, "A Simple Metal Lift-off Process," Solid-State Technology, *25*, 111–114 (February, 1982).

11. S. A. Evans, S. A. Morris, L. A. Arledge, Jr., J. O. Englade, and C. R. Fuller, "A 1-μm Bipolar VLSI Technology," IEEE Transactions on Electron Devices, *ED-27*, 1373–1379 (August, 1980).

12. Y. Sasaki, O. Ozawa, and S. Kameyama, "Application of $MoSi_2$ to the Double-Level Interconnections of I^2L Circuits," IEEE Transactions on Electron Devices, *ED-27*, 1385–1389 (August, 1980).

13. R. A. Larsen, "A Silicon and Aluminum Dynamic Memory Technology," IBM Journal of Research & Development, *24*, 268–282 (May, 1980).

14. J. M. Mikkelson, L. A. Hall, A. K. Malhotra, S. D. Seccombe, and M. S. Wilson, "An NMOS VLSI Process for Fabrication of a 32b CPU Chip," IEEE ISSCC Digest, *24*, 106–107 (February, 1981).

15. P. B. Ghate, "Multilevel Interconnection Technology," IEEE IEDM Digest, 126–129 (December, 1984).

PROBLEMS

7.1 (a) What is the sheet resistance of a 1-μm-thick aluminum-copper-silicon line with a resistivity of 3.2 μohm-cm?

(b) What would be the resistance of a line 500 μm long and 10 μm wide?

(c) What is the capacitance of this line to the substrate if it is on an oxide which is 1 μm thick? (Assume that you can use the parallel-plate capacitance formula).

(d) What is the *RC* product associated with this 500-μm line?

7.2 (a) Repeat Problem 7.1 for a polysilicon line with a resistivity of 500 μohm-cm.

(b) Repeat Problem 7.1 for a titanium silicide line with a resistivity of 25 μohm-cm.

7.3 (a) Compute estimates of the sheet resistance of shallow arsenic and boron diffusions by assuming uniformly doped rectangular regions with the maximum achievable electrically active impurity concentrations (see Fig. 4.6). Use hole and electron mobilities of 75 and 100 cm^2/V-sec, respectively, and a depth of 0.25 μm.

(b) Compare your answers with those for diffused lines obtained from Figs. 4.6 and 4.16. Use the maximum possible electrically active concentration for the boron and arsenic surface concentrations.

7.4 Suppose that a 500 $\times$ 15 μm aluminum line makes contact with silicon through a 10 $\times$ 10 μm contact window as shown in Fig. P7.4. The aluminum is 1 μm thick and is annealed at 450 °C for 30 min. Assume that the silicon will saturate the aluminum up to a distance $\sqrt{Dt}$ from the contact. D is the diffusion coefficient of silicon in aluminum which follows an Arrhenius relationship with $D = 0.04$ cm^2/sec and $E_A = 0.92$ eV. Assume that silicon is absorbed uniformly through the contact and that the density of aluminum and silicon is the same. How deep will the aluminum penetration into the silicon be?

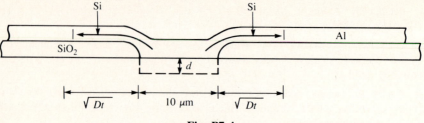

Fig. P7.4

7.5 A certain process forms aluminum contacts to n^+ silicon through a 10×10 μm contact window resulting in a contact resistance of 0.5 ohms.

(a) What is the specific contact resistivity for this contact?

(b) What will the contact resistance be if the contact windows are reduced to 1×1 μm? Does this seem acceptable for a VLSI process?

7.6 Electromigration failures depend exponentially on temperature.

(a) What is the ratio of the MTFs of identical aluminum conductors operating at the same current density at 300 K and 400 K?

(b) At 77 K, (liquid-nitrogen temperature) and 400 K? Use $E_A = 0.5$ eV.

7.7 An n^+ diffusion is used for interconnection. The surface concentration of the diffusion is $4 \times 10^{19}/\text{cm}^3$ and the junction depth is 4 μm. The diffusion is formed in a p-type wafer with a background concentration of $1 \times 10^{15}/\text{cm}^3$.

(a) What is the sheet resistance of this diffusion?

(b) Estimate the capacitance per unit length if the diffusion is 15 μm wide. Assume the rectangular geometry shown in Fig. P7.7 and use the step-junction-capacitance formula.

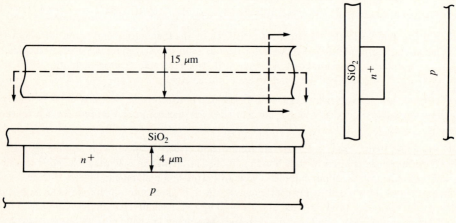

Fig. P7.7

7.8 What is the maximum current that may be allowed to flow in an aluminum conductor 1 μm thick and 4 μm wide if the current density must not exceed 5×10^5 A/cm²?

8 / Packaging and Yield

The low cost normally associated with integrated circuits results from mass production in which many wafers, each containing a large number of integrated-circuit dice, are all processed together. There may be tens to thousands of dice per wafer and 25 to 200 wafers per lot. After wafer processing is completed, however, the dice must be separated and assembled in packages that are easy to handle and to mount in electronic systems. The assembly operation involves a great deal of handwork and substantially increases the cost of the final product.

In this chapter we first present an overview of testing and die separation. Then we discuss IC assembly, including die attachment, wire bonding, and a survey of the various types of packages used with integrated circuits.

The ultimate cost of the integrated circuit is related to the total yield of assembled and tested devices. In the early stages of development of a new process or circuit, we are lucky if one functional die is found per wafer. Late in the life of a process with a mature circuit design, yields of 60 to 70% are not uncommon. A discussion of the dependence of yield on defect density and die size concludes this chapter.

8.1 TESTING

Following aluminum annealing and passivation-layer processing, each die on the wafer is tested for functionality. Special parametric test dice are placed at a number of sites on the wafer. At this stage, dc tests are used to verify that basic process parameters fall within acceptable limits. To perform the tests, a probe station lowers a ring of very fine, needle-sharp probes into contact with the aluminum pads on the test die. Test equipment is connected to the circuit through the probes and controlled by a computer system. If the wafer-screening operation shows that basic process and device parameters are within specification, functional testing of each die begins.

Under computer control, the probe station automatically steps across the wafer, performing functional testing at each die site. Defective dice are marked with a drop of ink. Later, when the dice are separated from the wafer, any die with an ink spot is discarded. It has become impossible to exhaustively test complex VLSI devices such as microprocessors. Instead, a great deal of computer time is used to find a minimum sequence of tests which can be used to indicate die functionality. At the wafer-probe stage, functional testing is primarily static in nature. High-speed dynamic testing is difficult to do through the probes, so parametric speed tests are usually performed after die packaging is complete.

The ratio of functional dice to total dice on the wafer gives the *yield* for each wafer. Yield is directly related to the ultimate cost of the completed integrated circuit and will be discussed more fully in Section 8.7.

8.2   DIE SEPARATION

Following initial functional testing, individual integrated-circuit dice must be separated from the wafer. In one method, the wafer is mounted on a holder and automatically scribed in both the *x* and *y* directions using a diamond-tipped scribe. Scribing borders of 75 to 250 μm are formed around the periphery of the dice during fabrication. These borders are left free of oxide and metal and are aligned with crystal planes if possible. In $\langle 100 \rangle$ wafers, natural cleavage planes exist perpendicular to the surface of the wafer in directions both parallel and perpendicular to the wafer flat. For $\langle 111 \rangle$ wafers, a vertical cleavage plane runs parallel to the wafer flat but not perpendicular to the wafer flat. This can lead to some separation and handling problems with $\langle 111 \rangle$ material.

Following scribing, the wafer is removed from the holder and placed upside down on a soft support. A roller applies pressure to the wafer, causing it to fracture along the scribe lines. Care is taken to ensure that the wafer cracks along the scribe lines to minimize die damage during separation, but there will always be some damage and loss of yield during the scribing and breaking steps.

Diamond saws are also widely used for die separation. A wafer is placed in a holder on a sticky sheet of Mylar. The saw can be used either to scribe the wafer or to cut completely through the wafer. Following separation, the dice remain attached to the Mylar film.

8.3   DIE ATTACHMENT

Visual inspection is used to sort out dice which may have been damaged during die separation, and the inked dice are also discarded. The next step in the assembly process is to mount the good dice in packages.

8.3.1   Epoxy Die Attachment

An epoxy cement may be used to attach the die to a package or "header." However, epoxy is a poor thermal conductor and an electrical insulator. Alumina can be mixed with the epoxy to increase its thermal conductivity, and gold- or silver-filled epoxies are used to reduce the thermal resistance of the epoxy bonding material and to provide a low-resistance electrical connection between the die and the package.

8.3.2   Eutectic Die Attachment

The gold-silicon eutectic point occurs at a temperature of 370 °C for a mixture of approximately 3.6% Si and 96.4% Au. Gold can be deposited on the back of the wafer prior to die separation or can be in the form of a thin alloy "preform" placed between the die and package. To form a eutectic bond, the die and package are heated to 390 to 420 °C, and pressure is applied to the die in conjunction with an ultrasonic scrubbing motion. Eutectic bonding is possible with a number of other metal-alloy systems, including gold-tin and gold-germanium. A solder attachment technique is also used with semiconductor power devices.

8.4   WIRE BONDING

Wire bonding is the most widely used method for making electrical connections between the die and the package. The bonding areas on the die are large, square pads, 100 to 125 μm on a side, located around the periphery of the die. Fine wires interconnect the aluminum bonding pads on the integrated-circuit die to the leads of the package. Thermocompression bonding was originally used with gold wire, and ultrasonic bonding is used with aluminum wire. A combination of the two, thermosonic bonding, is rapidly replacing thermocompression bonding.

8.4.1   Thermocompression Bonding

The thermocompression bonding technique, also called *nail head* or *ball bonding,* uses a combination of pressure and temperature to weld a fine gold wire to the aluminum bonding pads on the die and the gold plated leads of the package. Figure 8.1 shows the steps used in forming either a thermocompression or thermosonic bond.

A fine gold wire, 15 to 75 μm in diameter, is fed from a spool through a heated capillary. A small hydrogen torch or electric spark melts the end of the wire, forming a gold ball two to three times the diameter of the wire. Under either manual or computer control, the ball is positioned over the bonding pad, the capillary is lowered, and the ball deforms into a "nail head" as a result of the pressure and heat from the capillary.

Next, the capillary is raised and wire is fed from the spool as the tool is moved into position over the package. The second bond is a wedge bond produced by deforming the wire with the edge of the capillary. After formation of the second bond, the capillary is raised and the wire is broken near the edge of the bond. Because of the symmetry of the bonding head, the bonder may move in any direction following formation of the nailhead bond. An SEM micrograph of a gold-ball bond is shown in Fig. 8.2a.

A problem encountered in production of gold-aluminum bonds is formation of the "purple plague." Gold and aluminum react to form intermetallic compounds. One such compound, $AuAl_2$, is purple in color, and its appearance has been associated with faulty bonds. However, this compound is highly conductive. The actual culprit is a low-conductivity, tan-colored compound, Au_2Al, which is also present.

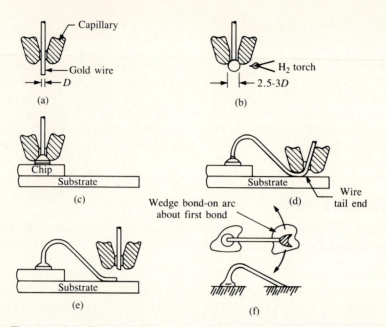

Fig. 8.1 Thermosonic ball-wedge bonding of a gold wire. (a) Gold wire in a capillary; (b) ball formation accomplished by passing a hydrogen torch over the end of the gold wire or by capacitance discharge; (c) bonding accomplished by simultaneously applying a vertical load on the ball while ultrasonically exciting the wire (the chip and substrate are heated to about 150 °C); (d) a wire loop and a wedge bond ready to be formed; (e) the wire is broken at the wedge bond; (f) the geometry of the ball-wedge bond that allows high-speed bonding. Because the wedge can be on an arc from the ball, the bond head or package table does not have to rotate to form the wedge bond. Reprinted with permission from Semiconductor International magazine, May 1982.[1] Copyright 1982 by Cahners Publishing Co., Des Plaines, IL.

During thermocompression bonding, the substrate is maintained at a temperature between 150 and 200 °C. The temperature at the bonding interface ranges from 280 to 350 °C, and significant formation of the Au-Al compounds can occur at these temperatures. Limiting the die temperature during the bonding process helps to prevent formation of the intermetallic compounds, and high-temperature processing and storage following bonding must be avoided.

In addition to the above problem, many epoxy materials cannot withstand the temperatures encountered in thermocompression bonding, and thermocompression bonding has been replaced by the ultrasonic and thermosonic bonding techniques discussed in the next two sections.

8.4.2 Ultrasonic Bonding

Oxidation of aluminum wire at high temperatures makes it difficult to form a good ball at the end of the wire. An alternative process called *ultrasonic bonding,* which forms the

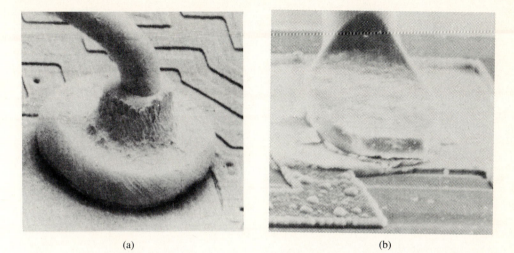

          (a)   (b)

Fig. 8.2 (a) An SEM micrograph of ball bond using 25-μm gold wire; (b) an SEM micrograph of 25-μm gold wire ultrasonically bonded to an aluminum pad. William C. Till/James T. Luxon, INTEGRATED CIRCUITS: Materials, Devices, & Fabrication, Copyright 1982, p. 360–361. Reprinted by permission of Prentice-Hall, Inc., Englewood Cliffs, New Jersey.

bond through a combination of pressure and rapid mechanical vibration, is used with aluminum wire. Aluminum wire is fed from a spool through a hole in the ultrasonic bonding tool, as shown in Fig. 8.3. To form a bond, the bonding tool is lowered over the bonding position, and ultrasonic vibration at 20 to 60 kHz causes the metal to deform and flow easily under pressure at room temperature. Vibration also breaks through the oxide film that is always present on aluminum and results in formation of a clean, strong weld.

    As the tool is raised after forming the second bond, a clamp is engaged which pulls and breaks the wire at a weak point just beyond the bond. In order to maintain proper wire alignment in the bonding tool, the ultrasonic bonder can move only in a front-to-back motion between the first and second bonds, and the package must be rotated 90° to permit complete bonding of rectangular dice.

8.4.3 Thermosonic Bonding

Thermosonic bonding combines the best properties of ultrasonic and thermocompression bonding. The bonding procedure is the same as in thermocompression bonding, except that the substrate is maintained at a temperature of approximately 150 °C. Ultrasonic vibration causes the metal to flow under pressure and form a strong weld. The symmetrical bonding tool permits movement in any direction following the nail-head bond. Thermosonic bonding can be easily automated, and computer-controlled thermosonic bonders can produce five to ten bonds per second.

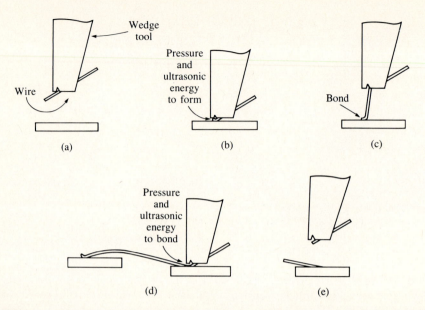

Fig. 8.3 (a) In ultrasonic bonding, the tool guides wire to the package terminal; (b) pressure and ultrasonic energy form the bond; (c) and (d) the tool feeds out wire and repositions itself above the IC chip. The tool lowers and ultrasonically forms the second bond; (e) tool lifts, breaking the wire at the bond. Reprinted with permission from Circuits Manufacturing, January, 1980. Copyright Morgan-Grampian 1980.

8.5   PACKAGES

Integrated-circuit dice can be mounted in a wide array of packages. In this section we discuss the round "TO"-style packages, dual-in-line (DIP) packages, the pin-grid array (PGA), the leadless chip carrier (LCC), and packages used for surface mounting. Later in this chapter we will look at flip-chip mounting and tape-automated bonding.

8.5.1   Circular TO-Style Packages

Figure 8.4a shows a round TO-type package which was one of the earliest IC packages. Different configurations of this package are available with 4 to 48 pins. The silicon die is attached to the center of the gold-plated header. Wire bonds connect the pads on the die to Kovar lead posts which protrude through the header and glass seal. Kovar is an iron-nickel alloy designed to have the same coefficient of thermal expansion as the glass seal. A metal cap is welded in place after die attachment and wire bonding.

8.5.2   Dual-in-Line Packages (DIPs)

The dual-in-line package shown in Fig. 8.4b is extremely popular because of its low cost and ease of use. Plastic and epoxy DIPs are the least expensive packages and are

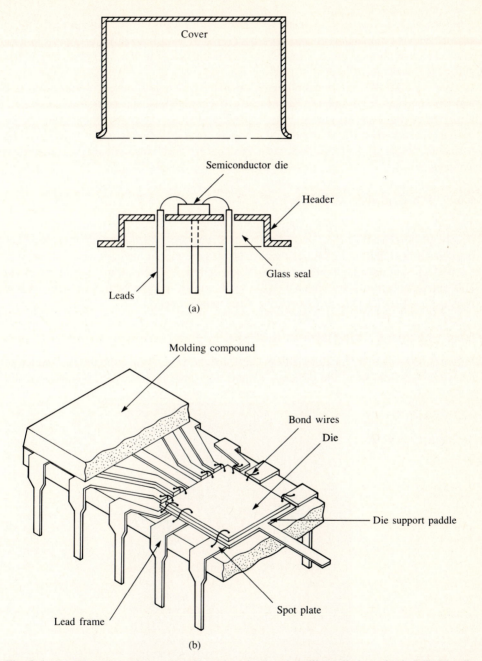

Fig. 8.4 (a) TO-style package; (b) ball- and wedge-bonded silicon die in a plastic DIP. The die support paddle may be connected to one of the external leads. For most commercial products, only the die paddle and the wedge-bond pads are selectively plated. The external leads are solder plated or dipped after package molding. Copyright, 1981, IEEE. Reprinted with permission from ref. [3].

available with as few as four leads to more than 80 leads. In the postmolded DIP, a silicon die is first mounted on and wire-bonded to a metallic lead frame. Epoxy is then molded around both the die and the frame. As a result, the silicon die becomes an integral part of the package.

In a ceramic DIP, the die is mounted in a cavity on a gold-plated ceramic substrate and wire-bonded to gold-plated Kovar leads. A ceramic or metal lid is used to seal the top of the cavity. Ceramic packages are considerably more expensive than plastic and are designed for use over a wider temperature range. In addition, ceramic packages may be hermetically sealed. A premolded plastic package similar to the ceramic package is also available.

8.5.3 Pin-Grid Arrays (PGAs)

The DIP package is satisfactory for packaging integrated-circuit dice with up to approximately 80 pins. The *pin-grid array* of Fig. 8.5 provides a package with a much higher pin density than that of the DIP package. The pins are placed in a regular *x-y* array, and the package can have hundreds of pins. Wire bonding is still used to connect the die to gold interconnection lines which fan out to the array of pins. Other versions of this package use the flip-chip process (to be discussed in Section 8.6).

8.5.4 Leadless Chip Carriers (LCCs)

Figure 8.6 shows two types of leadless chip carriers. In each case the die is mounted in a cavity in the middle of the package. Connections are made between the package and

Fig. 8.5   An example of a PGA in which the chip faces upward in the cavity.

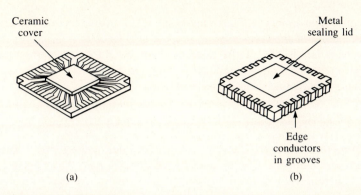

Ceramic
cover

Metal
sealing lid

Edge
conductors
in grooves

(a)   (b)

Fig. 8.6   (a) Ceramic leadless chip carriers with top connections; (b) LCC with edge connections in grooves on the sides of the package.

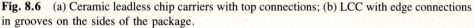

die using wire bonding, and the cavity is sealed with a cap of metal, ceramic, or epoxy. The package in Fig. 8.6a has contact pads only on the top surface of the chip carrier. The chip carrier is pressed tightly against contact fingers in a socket mounted on a printed-circuit board. Another type of LCC is shown in Fig. 8.6b. Conductors are formed in grooves in the edges of the chip carrier and are again pressed tightly against contact pins in a socket on the next level of packaging.

8.5.5 Packages for Surface Mounting

The TO-, DIP-, and PGA-style packages are made for mounting in holes fabricated in printed-circuit boards. New "surface-mount" packages have recently been developed. The "gull-wing" package, shown in Fig. 8.7a, has short lead stubs bent away from the package, whereas the leads of the "J"-style package of Fig. 8.7b are bent back underneath the package. Both styles permit soldering of the package directly to the surface of a printed-circuit board or hybrid package. The leadless chip carriers described in the previous section are also available with leads added for surface mounting.

8.6 FLIP-CHIP AND TAPE-AUTOMATED-BONDING PROCESSES

As can be envisioned from the above discussion, the die-mounting and wire-bonding processes involve a large number of manual operations and are therefore quite expensive.

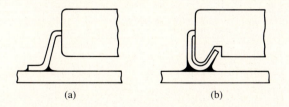

(a)   (b)

Fig. 8.7   The (a) gull-wing and (b) J-lead surface-mount packages.

In fact, the cost of assembly and test may be many times the cost of a small die. The one-at-a-time nature of the wire-bonding process also leads to reduced reliability, and failure of wire bonds is one of the most common reliability problems in integrated circuits. The flip-chip and tape-automated-bonding processes were developed to permit batch fabrication of die-to-package interconnections.

8.6.1 Flip-Chip Processing

The *flip-chip* mounting process was developed at IBM during the 1960s.[4] The first step is to form a solder ball (Fig. 8.8) on top of each bonding pad. A sandwich of Cr, Cu, and Au is sequentially evaporated through a mask to form a cap over each of the aluminum bonding pads. Chrome and copper provide a barrier and a good contact to the aluminum pad. Gold adheres well to chrome and acts as an oxidation barrier prior to solder deposition. Lead-tin solder is evaporated through a mask onto the Au-Cu-Cr cap, occupying an area slightly larger than the cap. The die is heated, causing the solder to recede from the oxide surface and form a solder ball on top of the Au-Cu-Cr bonding-pad cap.

After testing and separation, the dice are placed face down on a ceramic substrate. Temperature is increased, causing the solder to reflow, and the die is bonded directly to the interconnections on the substrate. Solder balls provide functions of both electrical interconnection and die attachment. Hundreds of bonds can be formed simultaneously using this technique, and bonding pads may be placed anywhere on the surface of the die, rather than just around the edge. In addition, the bond between the die and the substrate

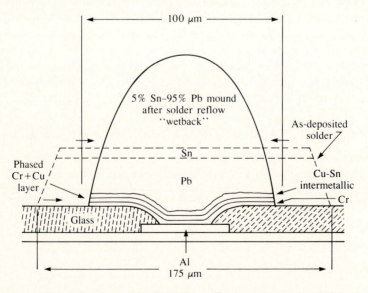

Fig. 8.8 Cross section through a solder ball after reflowing. Copyright 1969 by International Business Machines Corporation; reprinted with permission from ref. [4].

is very short. The main disadvantages of this technique are the additional processing complexity, the higher thermal resistance between die and substrate, and the inability to visually inspect the completed solder joints.

8.6.2   The Tape-Automated-Bonding (TAB) Process

In *tape-automated bonding,* dice are attached to copper leads supported by a tape similar to 35-mm film. The film is initially coated with copper, and the leads are defined by lithography and etching. The lead pattern may contain hundreds of connections.

Die attachment requires a process similar in concept to the solder-ball technology discussed earlier. Gold bumps are formed on either the die or the tape and are used to bond the die to the leads on the tape. Figure 8.9 outlines the steps used to form a gold bump on a bonding pad.[5] A multilayer metal sandwich is deposited over the passivation oxide. Next, a relatively thick layer of photoresist is deposited, and windows are opened above the bonding pads. Electroplating is used to fill the openings with gold. The photoresist is removed, and the thin metal sandwich is etched away using wet or dry etching. The final result is a 25-μm-high gold bump standing above each pad. As in the flip-chip approach, bonding sites may be anywhere on the die.

The mounting process aligns the tape over the die, as in Fig. 8.10. A heated bonding head presses the tape against the die, forming thermocompression bonds. In a production process, a new die is brought under the bonding head and the tape indexes automatically to the next lead site.

TAB-mounted parts offer the advantage that they can be functionally tested and "burned-in" once the dice are attached to the film. In addition, the IC passivation layer and gold bump completely seal the semiconductor surface.

8.7   YIELD

The manufacturer of integrated circuits is ultimately interested in how many finished chips will be available for sale. A substantial fraction of the dice on a given wafer will not be functional when they are tested at the wafer-probe step at the end of the process. Additional dice will be lost during the die separation and packaging operations, and a number of the packaged devices will fail final testing.

As mentioned earlier, the cost of packaging and testing is substantial and may be the dominant factor in the manufacturing cost of small die. For a large die with low yield, the manufacturing cost will be dominated by the wafer processing cost. A great deal of time has been spent attempting to model wafer yield associated with integrated-circuit processes. Wafer yield is related to the complexity of the process and is strongly dependent on the area of the integrated-circuit die.

8.7.1   Uniform Defect Densities

One can visualize how die area affects yield by looking at the wafer in Fig. 8.11, which has 120 die sites. The dots represent randomly distributed defects which have caused a

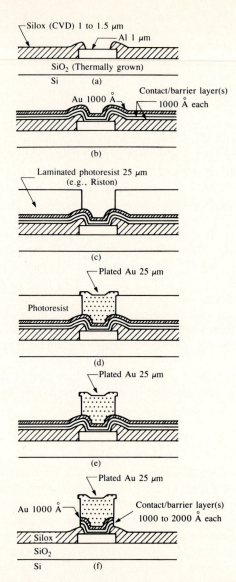

Fig. 8.9 Process sequence for making gold bumps on aluminum metallurgy devices. (a) The wafer is cleaned and sputter-etched; (b) a contact/barrier layer (which also serves as a conductive film for electroplating) is sputter-deposited with a layer of gold for oxidation protection; (c) a thick-film photoresist (25 μm) is laminated and developed; (d) gold is electroplated to a height of approximately 25 μm to form the bumps; (e) the resist is stripped; (f) the sputter-deposited conductive film is removed chemically or by back sputtering. Reprinted with permission of Solid State Technology, published by Technical Publishing, a company of Dun & Bradstreet, from ref. [5].

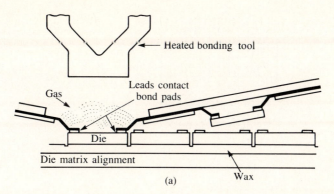

(a)

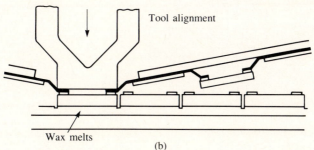

(b)

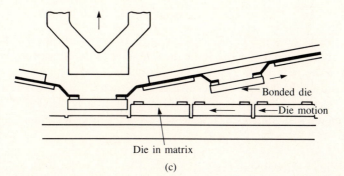

(c)

Fig. 8.10 Tape-automated-bonding procedure. (a) Preformed leads of film are lowered into position and aligned above bonding pads on the die, which is held in place with a wax; (b) bonding tool descends and forms bond with pressure and heat; heat melts the wax, releasing the die; (c) tool and film are raised, lifting the bonded die clear so a new die can be moved into position and the process can be repeated. Reprinted with permission from Small Precision Tools Bonding Handbook. Copyright 1976.

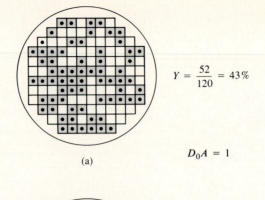

$$Y = \frac{52}{120} = 43\%$$

$$D_0 A = 1$$

(a)

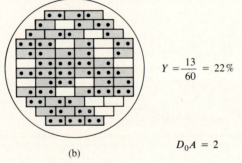

$$Y = \frac{13}{60} = 22\%$$

$$D_0 A = 2$$

(b)

Fig. 8.11 Illustrations of wafers, showing effect of die size on yield. Dots indicate the presence of a defective die location. (a) For a particular die size the yield is 43%; (b) if the die size were doubled, the yield would be only 22%.

die to fail testing at the wafer-probe step. In Fig. 8.11a there are 52 good dice out of the total of 120, giving a yield of 43%. If the die size were twice as large, as in Fig. 8.11b, the yield would be reduced to 22% for this particular wafer.

An estimate of the yield of good dice can be found from a classical problem in probability theory in which n defects are randomly placed in N die sites. The probability P_k that a given die site contains exactly k defects is given by the binomial distribution:

$$P_k = \frac{n!}{k!(n-k)!} N^{-n}(N-1)^{n-k} \tag{8.1}$$

For large n and N, eq. (8.1) can be approximated by the Poisson distribution:

$$P_k = \frac{\lambda^k}{k!} \exp(-\lambda) \tag{8.2}$$

in which $\lambda = n/N$. The yield is given by the probability that a die is found with no defects,

$$Y = P_0 = \exp(-\lambda) \tag{8.3}$$

The area of the wafer is equal to NA, where A is the area of one die. The density of defects, D_0, is given by the total number of defects, n, divided by the area of all the chips, and the average number of defects per die, λ, is given by

$$\lambda = n/N = D_0 A \quad \text{for} \quad D_0 = n/NA \tag{8.4}$$

The yield based on the Poisson distribution then becomes

$$Y = \exp(-D_0 A) \tag{8.5}$$

This expression was used to predict early die yield but was found to give too low an estimate for large dice with $D_0 A > 1$. Eq. (8.5) implicitly assumes that the defect distribution is uniform across a given wafer and does not vary from wafer to wafer. However, it was quickly realized that these conditions are not realistic. Defect densities vary from wafer to wafer because of differences in handling and processing. On a given wafer, there are usually more defects around the edge of the wafer than in the center, and the defects tend to be found in clusters. These realizations led to investigation of nonuniform defect densities.

8.7.2 Nonuniform Defect Densities

Murphy[6] showed that the wafer yield for a nonuniform defect distribution can be calculated from

$$Y = \int_0^\infty \exp(-DA) f(D) \, dD \tag{8.6}$$

in which $f(D)$ is the probability density for D. He considered several possible distributions, as shown in Fig. 8.12. The impulse function in Fig. 8.12a represents the case in which the defect density is the same everywhere, and substituting it for $f(D)$ in eq. (8.6) yields eq. (8.5). The triangular distribution in Fig. 8.12b is a simple approximation to a Gaussian distribution function and allows some wafers to have very few defects and others to have up to $2D_0$ defects. Application of eq. (8.6) results in the following yield expression:

$$Y = \left[\frac{1 - \exp(-D_0 A)}{D_0 A} \right]^2 \tag{8.7}$$

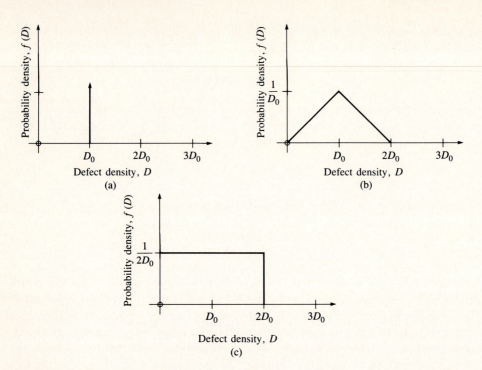

Fig. 8.12 Possible defect probability density functions. (a) Impulse, where every wafer has exactly the same number of defects; (b) a triangular approximation to a Gaussian density; (c) a uniform density function.

A uniform distribution of defect densities is modeled by $f(D)$ in Fig. 8.12c and predicts a yield of

$$Y = \left[\frac{1 - \exp(-2D_0 A)}{2D_0 A} \right]$$ (8.8)

More-complicated probability distributions have also been investigated, including the negative binomial and gamma distributions.[7, 8] These result in the yield expression in eq. (8.9):

$$Y = \left[1 + \frac{D_0 A}{\alpha} \right]^{-\alpha}$$ (8.9)

in which α represents a clustering parameter which ranges from 0.5 to 5.

Figure 8.13 plots the various yield functions versus $D_0 A$, the average number of defects in a die of area A. Yield estimates based on Poisson statistics are clearly much more pessimistic than those based on the other functions.

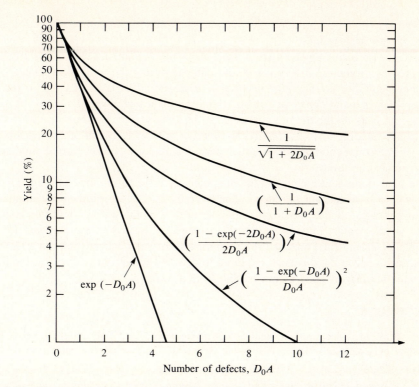

Fig. 8.13 Theoretical yield curves for different defect densities. See eqs. (8.5) through (8.9).

Example 8.1: A 150-mm wafer has a defect density of 10 defects/cm^2, and costs $200 to process. The cost of assembly and testing is $1.50 per die. **(a)** What is the total manufacturing cost for a 5 × 5 mm die in this process based on yield eq. (8.7)? (The number of square dice per wafer is given approximately by $N = \pi(R - S)^2/S^2$ where R is the wafer radius and S is the length of the side of the die.) **(b)** The market price for this part is $2.50. What must be the wafer yield needed for manufacturing cost to drop below the market price?

Solution: The area of the die is 0.25 cm, so the average number of defects per die is $D_0 A = 2.5$. Eq. (8.7) predicts a yield of 13.5%. The wafer has a radius of 75 mm and contains approximately 616 dice. So the average wafer will yield 83 good dice. The cost of the packaged dice will be $C = (\$200/83) + \$1.50 = \$3.91$. In order to get the cost to the market price requires $\$2.50 = (\$200/ND) + \$1.50$. We must get $ND = 200$ good dice per wafer to break even, corresponding to a yield of $Y = 200/616 = 0.325$ or more.

8.8 SUMMARY

Following the completion of processing, wafers are screened by checking various processing and device parameters using special test sites on the wafer. If the parameters are within proper limits, each die on the wafer is tested for functionality, and bad dice are marked with a drop of ink.

Next, the dice are separated from the wafer using a diamond saw or a scribe-and-break process. Some die loss is caused by damage during the separation process. The remaining good dice are mounted in ceramic or plastic DIPs, LCCs, PGAs, or surface-mount packages using epoxy or eutectic die-attachment techniques.

Bonding pads on the die are connected to leads on the package using ultrasonic or thermosonic bonding of 15 to 75 μm aluminum or gold wire. Batch-fabricated flip-chip and TAB interconnection processes which permit simultaneous formation of hundreds of bonds can also be used.

The final manufacturing cost of an integrated circuit is determined by the number of functional parts which are produced. The overall yield is the ratio of the number of working packaged dice to the original number of dice on the wafer. Yield loss is due to defects on the wafer, processing errors, damage during assembly, and lack of full functionality during final testing. The relationship between wafer yield and the size of an integrated-circuit die has been explored in detail. The larger the die size, the lower will be the number of good dice available from a wafer.

REFERENCES

[1] J. W. Stafford, "The Implications of Destructive Wire Bond Pull and Ball Bond Shear Testing on Gold Ball–Wedge Wire Bond Reliability," Semiconductor International, p. 82 (May, 1982).

[2] W. C. Till and J. T. Luxon, *Integrated Circuits: Materials, Devices and Fabrication,* Prentice-Hall, Englewood Cliffs, NJ, 1982.

[3] J. R. Howell, "Reliability Study of Plastic Encapsulated Copper Lead Frame Epoxy Die Attach Packaging System," Proceedings of the International Reliability Physics Symposium, p. 104–110, 1981.

[4] P. A. Totta and R. P. Sopher, "SLT Device Metallurgy and Its Monolithic Extension," IBM Journal of Research & Development, *13,* 226–238 (May, 1969).

[5] T. S. Liu, W. R. Rodgrigues de Miranda, and P. R. Zipperlin, "A Review of Wafer Bumping for Tape Automated Bonding," Solid State Technology, *23,* 71–76 (March, 1980).

[6] B. T. Murphy, "Cost-Size Optima of Monolithic Integrated Circuits," Proceedings of the IEEE, *52,* 1537–1545 (December, 1964).

[7] R. B. Seeds, "Yield and Cost Analysis of Bipolar LSI," IEEE IEDM Proceedings, p. 12 (October, 1967).

[8] C. H. Stapper, "On Yield, Fault Distributions, and Clustering of Particles," IBM Journal of Research & Development, *30,* 326–338 (May, 1986).

FURTHER READING

1. G. G. Harman and J. Albers, "The Ultrasonic Welding Mechanism as Applied to Aluminum- and Gold-Wire Bonding in Microelectronics," IEEE Transactions on Parts, Hybrids and Packaging, *PHP-13,* 406–412, (December, 1977).

2. K. I. Johnson, M. H. Scott, and D. A. Edson, "Ultrasonic Wire Welding–Part I: Wedge-Wedge Bonding of Aluminum Wires," Solid State Technology, *20,* 50–56 (March, 1977).

3. K. I. Johnson, M. H. Scott, and D. A. Edson, "Ultrasonic Wire Welding–Part II: Ball-Wedge Wire Welding," Solid State Technology, *20,* 91–95 (April, 1977).

4. C. Plough, D. Davis, and H. Lawler, "High Reliability Aluminum Wire Bonding," Proceedings of the Electronic Components Conference, p. 157–165 (1969).

5. N. Ahmed and J. J. Svitak, "Characterization of Gold-Gold Thermocompression Bonding," Proceedings of the Electronic Components Conference, p. 92–97 (1976).

6. L. S. Goldmann, "Geometric Optimization of Controlled Collapse Interconnections," IBM Journal of Research & Development, *13,* 251–265 (May, 1969).

7. K. C. Norris and A. H. Landzberg, "Reliability of Controlled Collapse Interconnections," IBM Journal of Research & Development, *13,* 266–271 (May, 1969).

8. J. E. Price, "A New Look at Yield of Integrated Circuits," Proceedings of the IEEE, *58,* 1290–1291 (August, 1970).

9. C. H. Stapper, Jr., "On a Composite Model to the IC Yield Problem," IEEE Journal of Solid-State Circuits, *SC-10,* 537–539 (December, 1975).

10. C. H. Stapper, "The Effects of Wafer to Wafer Defect Density Variations on Integrated Circuit Defect and Fault Distributions," IBM Journal of Research & Development, *29,* 87–97 (January, 1985).

PROBLEMS

8.1 Make a list of at least ten process or device parameters which could easily be monitored using a special test site on a wafer.

8.2 A simple microprocessor contains 115 flip-flops and hence 2^{115} possible states. If a tester can perform a new static test every 100 nsec, how many years will it take to test every state in the microprocessor chip? If the wafer has 100 dice, how long will it take to test the wafer?

8.3 Compare the four yield formulas for a large VLSI die in which $D_0 A = 10$ defects. Assume a clustering parameter of 1.0. How many good dice can we expect from 100- and 150-mm-diameter wafers using the different yield expressions? (The number of square dice per wafer can be estimated from $N = \pi (R - S)^2 / S^2$ where R is the radius of the wafer and S is the length of one side of the die.) Assume $S = 5$ mm.

8.4 What is the wafer yield for the defect map in Fig. 8.11 if the die is four times the size of that in Fig. 8.11a? What is the yield predicted by Poisson statistics? Assume the data from Fig. 8.11 is best represented by eq. (8.9). What value of clustering parameter best fits the data?

8.5 A new circuit design is estimated to require a die which is 5 mm $\times$ 8 mm and will be fabricated on a wafer 125 mm in diameter. The process is achieving a defect density of 10 defects/cm^2, and the wafer processing cost is $250.

(a) What will be the cost of the final product if testing and packaging adds $1.60 to the completed product?

(b) The circuit design could be partitioned into two chips rather than one, but each die will increase in area by 15% in order to accommodate additional pads and I/O circuitry. If the testing and packaging cost remains the same, what is the cost of the two-chip set? Base your answers on eq. (8.8). (See Problem 8.3 for the number of dice per wafer.)

8.6 (a) Repeat Problem 8.5 for a defect density of 5 defects/cm² and a wafer cost of $150.

(b) Repeat Problem 8.5 for a defect density of 5 defects/cm² and a wafer cost of $300.

8.7 A die has an area of 25 mm² and is being manufactured on a 100-mm-diameter wafer using a process rated at 2 defects/cm². A new process is being developed which allows the die area to be reduced by a factor of 2. However, because of the smaller feature sizes, the new process costs 30% more and is presently achieving only 10 defects/cm².

(a) Is it economical to switch to this new process?

(b) At what defect density does the cost of the new die equal the cost of the old die?

(c) Based on your judgment, would you recommend switching to the new process even if it is not now economical? Why?

(d) At what die size is the cost the same in either process? Use eq. (8.8) for this problem.

8.8 What is the limit of the yield distribution in eq. (8.9) as the clustering parameter approaches infinity?

8.9 Suppose that going from 100-mm wafers to 150-mm wafers changes the wafer processing cost from $150/wafer to $250/wafer, and the defect density remains constant at 10 defects/cm². What two die sizes give the same die cost? Use eq. (8.9) with a cluster factor of 2. Use a calculator or personal computer to find the answer by iteration.

8.10 What would be the die yield in Fig. 8.11b if the defect positions were the same but the die pattern was rotated by 90°? How many good dice with four times the area of that in Fig. 8.11a would now exist?

8.11 A Gaussian probability density function for defect density is given by

$$f(D) = \frac{2}{D_0\sqrt{\pi}} \exp - \left[\frac{2(D - D_0)}{D_0} \right]^2 \quad \text{for} \quad 0 \le D \le 2D_0 \quad \text{and 0 otherwise.}$$

Calculate the yield Y for various values of $D_0 A$ and compare your results to those of the triangular distribution given in eq. (8.7). (You may want to use a calculator or computer to perform the integration.)

8.12 The wafers shown in Fig. 8.11 actually have 120 defects placed randomly on the wafer. Obviously, some chips must have several defects. Use eq. (8.1) to predict how many dice will have exactly 1, 2, 3, 4, and 5 defects.

9 / MOS Process Integration

In Chapter 9 we explore a number of relationships between process and device design and circuit layout. Processes are usually developed to provide devices with the highest possible performance in a specific circuit application, and one must understand the circuit environment and its relation to device parameters and device layout.

In this chapter we look at a number of basic concerns in MOS process design, including channel-length control; layout ground rules and ground-rule design; source/drain breakdown and punch-through voltages; and threshold-voltage adjustment. Metal-gate technology is discussed first, and then the important advantages of self-aligned silicon-gate technologies are presented. Discussions of CMOS technology and the application of anisotropic etching to MOS devices complete the chapter.

9.1 BASIC MOS DEVICE CONSIDERATIONS

To explore the relationship between MOS process design and basic device behavior, we begin by discussing the static current-voltage relationship for the MOS transistor, as developed in Volume IV of this series.[1] The cross section of two metal-gate NMOS transistors is shown in Fig. 9.1. In the linear region of operation, the drain current is given by

$$I_D = \overline{\mu}_n C_0 (Z/L)(V_{GS} - V_T - V_{DS}/2)V_{DS} \qquad (9.1)$$

for $V_{GS} \geq V_T$ and $V_{DS} \leq V_{GS} - V_T$. $C_0 = K_0\varepsilon_0/X_0$ is the oxide capacitance per unit area, $\overline{\mu}_n$ is the average majority-carrier mobility in the inversion layer, and V_T is the threshold voltage.

One of the first specifications required is the circuit-power-supply voltages, which set the maximum value of V_{GS} and V_{DS} that the devices must withstand. Once this choice is made, the only variables in eq. (9.1) which a circuit designer may adjust are the width and length of the transistor. Thus, the circuit designer varies the circuit topology and horizontal geometry to achieve the desired circuit function.

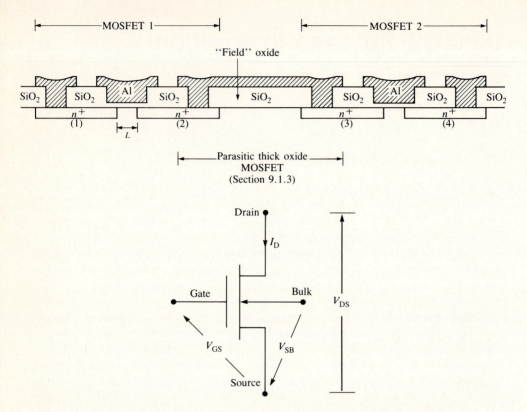

Fig. 9.1 (a) Cross section of an integrated circuit showing two adjacent NMOS transistors. A parasitic NMOS device is formed by the aluminum interconnection over the field oxide with diffused regions (2) and (3) acting as source and drain. (b) An NMOS transistor with gate-to-source (V_{GS}), drain-to-source (V_{DS}), and source-to-bulk (V_{SB}) voltages defined.

Other device parameters are fixed by the process designer, who must determine the process sequence, times, temperatures, etc., which ultimately determine the device structure and hence its characteristics. These include specifying the gate-oxide thickness, field-oxide thickness, substrate doping, and field and threshold-adjustment implantations. The process designer also supplies a set of "design rules" or "ground rules" which must be obeyed during circuit layout. These include minimum channel length and width, spacings between features on the same and different mask levels, and overlaps between features on different mask levels. A mask alignment sequence and tolerances must also be developed for the process.

9.1.1 Gate-Oxide Thickness

Current flow in the MOS transistor, for a given set of terminal voltages, is inversely proportional to the gate-oxide thickness. The gate oxide will generally be made as thin as possible, commensurate with oxide breakdown and reliability considerations. High-quality silicon dioxide will typically break down at electric fields of 5 to 10 MV/cm, corresponding to 50 to 100 V across a 100-nm oxide. Present processes are using oxide thicknesses between 20 and 100 nm. Below 10 nm, current starts to flow by tunneling, and the oxide begins to lose its insulating qualities. The choice of oxide thickness is also related to hot electron injection into the oxide, a problem beyond the scope of this text.[2–4]

9.1.2 Substrate Doping and Threshold Voltage

Threshold voltage is an important parameter which determines the gate voltage necessary to initiate conduction in the MOS device. The threshold voltage[1] for a device with a uniformly doped substrate is given by:

$$\text{NMOS: } V_T = \Phi_M - \chi - \frac{E_g}{2q} + |\Phi_F| + [\sqrt{2K_s\varepsilon_0 qN_B(2|\Phi_F| + V_{SB})}]/C_0 - Q_{tot}/C_0$$

$$(9.2)$$

$$\text{PMOS: } V = \Phi_M - \chi - \frac{E_g}{2q} - |\Phi_F| - [\sqrt{2K_s\varepsilon_0 qN_B(2|\Phi_F| - V_{BS})}]/C_0 - Q_{tot}/C_0$$

$$|\Phi_F| = (kT/q)\ln(N_B/n_i)$$

in which N_B is the substrate doping. $\Phi_M - \chi = -0.11$ for an aluminum gate, $\Phi_M - \chi = 0$ for an n^+-doped polysilicon gate, and $\Phi_M - \chi = +1.12$ for a p^+-doped polysilicon gate.

$Q_{tot}{}^+$ represents the total oxide and interface charge per cm^2 and adds a parallel shift of the curves in Fig. 9.2 to more negative values of V_T. This charge contribution to the threshold voltage had an extremely important influence on early MOS device fabrication. Q_{tot} tends to be positive, which makes the MOS transistor threshold more negative; n-channel transistors become depletion-mode devices ($V_T < 0$), whereas p-channel transistors remain enhancement-mode devices ($V_T < 0$). During early days of MOS technology, Q_{tot} was high, and the only successful MOS processing was done using PMOS technology. After the industry gained an understanding of the origin of oxide and interface charges, and following the advent of ion implantation, NMOS technology became dominant because of the mobility advantage of electrons over holes. Today, total charge levels have been reduced to less than 5×10^{10} charges/cm^2 in good MOS processes, and the oxide charge contribution to threshold voltage is minimal.

Substrate doping enters the threshold-voltage expression through both the $|\Phi_F|$ term and the square-root term. A plot of threshold voltage versus substrate doping for n- and p-channel, n^+ polysilicon-gate devices with 50-nm gate oxides is given in Fig. 9.2 for $Q_{tot} = 0$. The choice of substrate doping is complicated by other considerations including

---

$^+Q_{tot} = Q_F + Q_{IT} + \gamma_M Q_M$

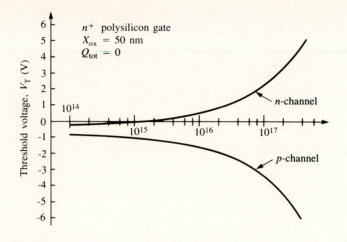

Fig. 9.2 Threshold voltages for *n*- and *p*-channel polysilicon-gate transistors with 50-nm gate oxides, calculated from eq. (9.2).

drain-to-substrate breakdown voltage, drain-to-source punch-through voltage, source-to-substrate and drain-to-substrate capacitances, and substrate sensitivity or body effect.

The source and drain regions are usually heavily doped to minimize their resistance and are essentially one-sided junctions in which the depletion region extends entirely into the substrate. Figure 9.3a gives the breakdown voltage of a one-sided *pn* junction as a function of the doping concentration on the lightly doped side of the junction.[5] Junction breakdown voltage decreases as doping level increases. Breakdown voltage is also a function of the radius of curvature of the junction space-charge region. Junction curvature enhances the electric field in the curved region of the depletion layer and reduces the breakdown voltage below that predicted by one-dimensional junction theory. A rectangular diffused area has regions with both cylindrical and spherical curvature, as shown in Fig. 9.3b.

Punch-through occurs when the drain depletion region contacts the source depletion region, and substrate doping must be chosen to prevent the merging of these depletion regions when the MOSFET is off. Punch-through will not occur if the channel length exceeds the sum of the depletion-layer widths of the source-to-substrate and drain-to-substrate junctions. For a transistor used as a load device in a logic circuit, the source-to-substrate and drain-to-substrate junctions must both support a voltage equal to the drain supply voltage plus the substrate supply voltage. The depletion-layer widths can be estimated using the formula for the width of a one-sided step junction:

$$W = \sqrt{(2K_s\varepsilon_0(V_A + \Phi_{bi}))/qN_B}$$
$$\Phi_{bi} = 0.56 + (kT/q)\ln(N_B/n_i) \tag{9.3}$$

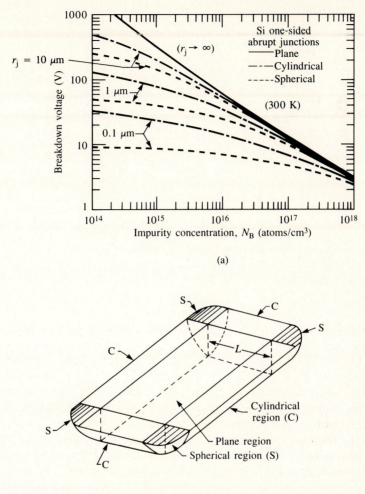

(a)

(b)

Fig. 9.3 (a) Abrupt *pn* junction breakdown voltage versus impurity concentration on the lightly doped side of the side of the junction for both cylindrical and spherical structures. r_j is the radius of curvature. (b) Formation of cylindrical and spherical regions by diffusion through a rectangular window. Copyright, 1985, John Wiley & Sons, Inc. Reprinted with permission from ref. [5].

where V_A is the total applied voltage and Φ_{bi} is the built-in potential of the junction. If the channel length is greater than $2W$, punch-through should not occur. Figure 9.4 gives the depletion-layer width of *pn* junctions as a function of doping and applied voltage. Punch-through is not a limiting factor for most doping levels, except for very short-channel transistors.

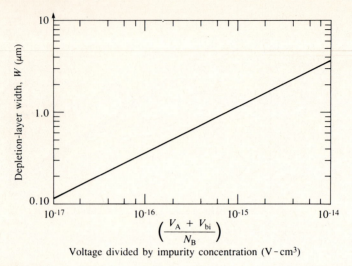

Fig. 9.4   Depletion-layer width of a one-sided step junction as a function of doping and applied voltage calculated from eq. (9.3).

The capacitance per unit area associated with a diffused junction is given by the parallel-plate capacitance formula with a plate spacing of W:

$$C_j = K_s \varepsilon_0 / W$$

The larger the doping, the larger the capacitance. Zero bias and a doping concentration of $10^{16}/cm^3$ result in a junction capacitance of approximately 10 nf/cm². Eq. (9.2) shows that the threshold voltage depends on the source-to-substrate voltage, V_{SB}. This variation is known as "substrate sensitivity" or "body effect," and it becomes worse as the substrate doping level increases.

From the above discussion, one can see that there are tradeoffs involved in the choice of substrate doping. Substrate doping is directly related to threshold voltage. It is desirable to reduce substrate doping to minimize junction capacitance and substrate sensitivity and to maximize breakdown voltage. Mobility also tends to be higher for lower doping levels. On the other hand, a heavily doped substrate will increase the punch-through voltage.

9.1.3   Threshold Adjustment

Ion implantation is routinely used to separate threshold-voltage design from the other factors involved in the choice of substrate doping. Substrate doping can be chosen based on a combination of breakdown, punch-through, capacitance, and substrate sensitivity considerations, and the threshold voltage is then adjusted to the desired value by adding

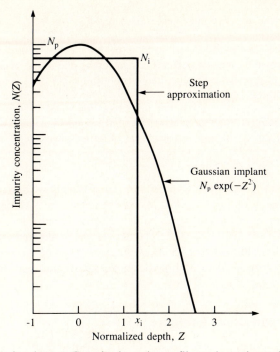

Fig. 9.5 Step approximation to a Gaussian impurity profile used to estimate the threshold-voltage shift achieved using ion implantation.

a shallow ion-implantation step to the process. Figure 9.5 shows a step approximation to an implanted profile used to adjust the impurity concentration near the surface. These additional impurities cause a shift in threshold voltage given approximately by

$$\Delta V_{\rm T} = (1/C_0)\,(qQ_{\rm i})\,(1 - x_{\rm i}/2x_{\rm d}), \qquad x_{\rm i} \ll x_{\rm d}, \qquad x_{\rm d} = \sqrt{qN_{\rm B}/4K_{\rm s}\varepsilon_0|\Phi_{\rm F}|}$$

$$(9.4)$$

where $Q_{\rm i} = x_{\rm i}N_{\rm i}$ represents the implanted dose and $x_{\rm d}$ represents the depletion-layer width beneath the gate. For shallow implants, the threshold-voltage shift is approximately proportional to the implanted dose. The threshold-voltage shift is positive for acceptor impurities and negative for donor impurities.

Example 9.1: An NMOS transistor with an n^+ polysilicon gate is fabricated with a 25-nm gate oxide, a substrate doping of $3 \times 10^{15}/{\rm cm}^3$, and source/drain junction depths of 3 μm. Determine the threshold voltage and drain-to-substrate breakdown voltages for this device. What is the punch-through voltage for a channel length of 4 μm if the substrate bias is -3 V? A shallow boron implantation is to be used to adjust the threshold to 1.0 V. What is the dose of this implant? Assume $V_{\rm SB} = 0$ and $Q_{\rm tot} = 0$.

Solution: For the n^+ polysilicon-gate transistor, $\Phi_M - \chi - E_g/2q = -0.56$ V and $|\Phi_F| = 0.33$ volts (for $n_i = 1 \times 10^{10}/cm^3$ and $kT/q = 0.026$ V). For $V_{SB} = 0$, the threshold voltage expression yields $V_T = -0.56 + 0.33 + 0.19$ V $= 0.04$ V. Interpolating Fig. 9.3 for spherical breakdown with a substrate doping of $3 \times 10^{15}/cm^3$ and a radius of curvature of 3 μm gives an estimated drain-to-substrate breakdown voltage of 60 V. To estimate the punch-through voltage, we use eq. (9.3) with $2W = 4$ μm and $V_A = V_D + 3$, where V_D is the drain voltage. Evaluating this expression yields $V_D = 5.4$ V.

For a shallow implant, the threshold-voltage shift is approximately $\Delta V_T = q\Delta Q/C_0$. A voltage shift of 1.04 V with an oxide thickness of 25 nm yields $\Delta Q = 9.0 \times 10^{11}/cm^2$.

NMOS depletion-mode ($V_T < 0$) transistors are routinely used in processes designed for high-performance logic applications. In order to reduce the NMOS threshold voltage, n-type impurities are implanted to form a built-in channel connecting the source and drain regions of the transistor, as in Fig. 9.6. The device characteristics of a depletion-mode transistor are similar, although not identical, to those of an enhancement-mode NMOS transistor, and the dose needed to shift the threshold voltage may be estimated using eq. (9.4).

9.1.4   Field-Region Considerations

The region between the two transistors in Fig. 9.1 is called the *field* region and must be designed to provide isolation between adjacent MOS devices. Several factors must be considered. The metal line over the field region can act as the gate of a "parasitic NMOS transistor" with diffused regions (2) and (3) acting as its source and drain. In order to ensure that this parasitic device is never turned on, the magnitude of the threshold voltage in this region must be much higher than that in the normal gate region. Referring to eq. (9.2), the threshold voltage may be made higher by increasing the oxide thickness

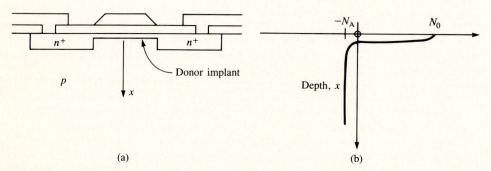

Fig. 9.6   (a) Formation of a depletion-mode NMOS transistor using a shallow ion-implanted layer; (b) net impurity profile under the gate of the depletion-mode MOSFET.

in the field region and by increasing the doping below the field oxide. The "field oxide" is typically made three to ten times thicker than the gate oxide of the transistors.

Another problem occurs for NMOS transistors. The substrate for NMOS transistors is *p*-type, usually doped with boron. We know that thermal oxidation results in depletion of boron from the surface of the silicon, and looking at eq. (9.2) we see that boron depletion will lower the threshold voltage of the transistors in the field region. A field implant step is often added to modern processes to increase the threshold voltage and compensate for the boron depletion during field-oxide growth.

For PMOS devices, the substrate is typically doped with phosphorus. During oxidation, phosphorus pileup at the surface tends to increase the threshold voltage in the field region. Thus phosphorus pileup helps to keep the parasitic field devices turned off.

One must also ensure that parasitic conduction does not occur between two adjacent devices due to punch-through. The source and drain diffusions of each transistor must be spaced far enough from the source and drain diffusions of the other transistors to ensure that the depletion regions do not merge together. The spacing between adjacent transistors must be greater than twice the maximum depletion-layer width.

9.2 MOS TRANSISTOR LAYOUT AND DESIGN RULES

Design of the layout for transistors and circuits is constrained by a set of rules called the "design rules" or "ground rules." These rules are technology-specific and specify minimum sizes, spacings, and overlaps for the various shapes that define transistors. Processes are designed around a "minimum feature size," which is the width of the smallest line or space that can be reliably transferred to the surface of the wafer using a given generation of lithography.

To produce a basic set of ground rules, we must also know the maximum misalignment which can occur between two mask levels. Figure 9.7a shows the nominal

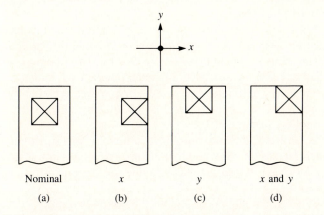

Nominal	x	y	x and y
(a)	(b)	(c)	(d)

Fig. 9.7 (a) Nominal alignment of the contact and metal masks; (b) worst-case misalignment in the *x*-direction, (c) in the *y*-direction, and (d) in both directions.

position of a metal line aligned over a contact window. The metal overlaps the contact window by at least one "alignment tolerance" in all directions. During the fabrication process, the alignment will not be perfect, and the actual structure may have misalignment in both the x and y directions. Figures 9.7b through d show the result of worst-case misalignment of the patterns in the x, y, and both directions simultaneously. Our set of design rules will assume that this "alignment tolerance" is the same in both directions.

9.2.1   Metal-Gate Transistor Layout

Figure 9.8 shows the process sequence for a basic metal-gate process. The first mask defines the position of the source and drain diffusions. Following diffusion, the second mask is used to define a window for growth of the thin gate oxide. The third and fourth masks delineate the contact openings and metal pattern. The metal-gate mask sequence, omitting the final passivation layer mask, is as follows:

1. Source/drain diffusion mask   First mask
2. Thin oxide mask   Align to level 1
3. Contact window mask   Align to level 1
4. Metal mask   Align to level 2

An alignment sequence must be specified in order to properly account for alignment tolerances in the ground rules. In this metal-gate example, mask levels two and three are aligned to the first level, and level four is aligned to level two.

We will first look at a set of design rules for metal-gate transistors similar in concept to the rules developed by Mead and Conway.[6] These ground rules were designed to permit easy movement of a design from one generation of technology to another by simply changing the size of a single parameter, λ. In order to achieve this goal, the rules are quite loose in terms of level-to-level alignment tolerance. We will explore tighter ground rules later in this chapter.

A set of metal-gate rules is shown in Fig. 9.9. The minimum feature size $F = 2\lambda$, and the alignment tolerance $T = \lambda$. The parameter λ could be 5 μm, 2 μm, or 1 μm, for example. Transistors designed using our ground rules will fail to operate properly if the misalignment exceeds the specified alignment tolerance T.

On the metal level, minimum line widths and spaces are equal to 2λ. In some processes, the metal widths are made larger because the metal level encounters the most mountainous topology of any level.

On the diffusion level, the minimum linewidth is 2λ. The minimum space between diffusions is increased to 3λ to ensure that the depletion layers of adjacent lines do not merge together. However, the spacing between the source/drain diffusions of a transistor may be 2λ.

In this set of rules, the alignment tolerance between two mask levels is assumed to be 1λ, which represents the maximum shift of one level away from its nominal position, relative to the level to which it is being aligned. A 1λ shift can occur in both the x and y directions.

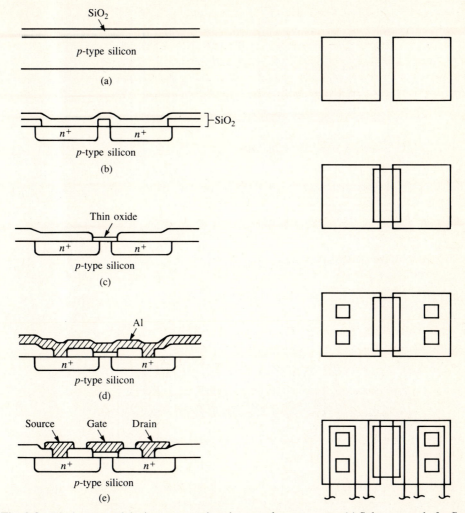

Fig. 9.8 Mask steps and device cross sections in a metal-gate process. (a) Substrate ready for first mask step; (b) substrate following source/drain diffusion and oxide regrowth, (c) following gate-oxide growth, (d) following contact window mask and aluminum deposition, and (e) following metal delineation.

Square contacts are a minimum feature size of 2λ in each dimension. It is normal practice to ensure that the contact is completely covered by metal even for worst-case alignment. Depending on the alignment sequence, a 1λ or 2λ metal border will be required around the contact window. Likewise, a contact window must be completely surrounded by a 1λ or 2λ border of the diffused region beneath the contact.

For our metal-gate transistors, the thin oxide region will be aligned to diffusion, so it requires a 1λ overlap over the source/drain diffusions in the length direction. The

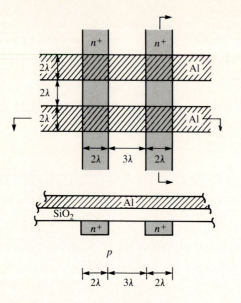
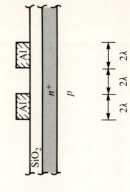
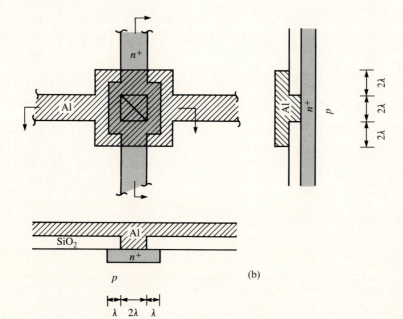

Fig. 9.9 A simple "λ-based" set of "design rules" or "ground rules" based on an alignment sequence in which levels 2 and 3 are aligned to level 1 and level 4 is aligned to level 2. (a) Rules for metal and diffused interconnection lines; (b) rules for contacts between metal and diffusion.

source/drain regions must also extend past the thin oxide by at least 1λ in the width direction. Contacts must be inside the diffusions by 1λ. The metal level is aligned to the thin oxide level, whereas the contacts are aligned to the diffusion level. A worst-case layout therefore requires a 2λ border of metal around contact windows but only a 1λ border around the thin oxide regions.

Figure 9.10 shows the horizontal layout and vertical cross section of a minimum-size NMOS metal-gate transistor with $Z/L = 10\lambda/2\lambda = 5/1$ at the mask level. The two diffusions are spaced by a minimum feature size of 2λ. Thin oxide must overlap the diffusions by 1λ in the length direction and underlap the diffusions by 1λ in the width direction. Metal must overlap thin oxide by 1λ. Accumulated alignment tolerances cause the minimum width of the gate metal to be 6λ. The spacing between metal lines must be 2λ. The metal over the contact holes must be 8λ wide because of the alignment sequence used, and the contact hole must be 1λ inside the edge of the diffusion. The resulting minimum transistor is 26λ in the length direction and 16λ in the width direction.

A new design rule has been introduced into this layout. The gate metal is spaced 1λ from the diffusion to prevent the edge of a metal line from falling directly on top of the edge of the diffusion in the nominal layout.

Several observations can be made by looking at this structure. First, note that the transistor is $416\lambda^2$ in total area, whereas the active channel area of the device is $20\lambda^2$! The rest of the area is required in order to make contacts to the various regions, within the constraints of the minimum feature size and alignment tolerance rules. Second, there is a substantial area of thin and thick oxide in which the gate metal overlaps the source and drain regions of the transistor. This increases the gate-to-source and gate-to-drain capacitance of the transistor. In this metal-gate transistor layout, the channel is defined by the junction edges in the length direction and by the thin oxide region in the width direction.

It should also be noted that there are several small contact windows in the source and drain regions. The usual practice is to make all the contact windows the same size throughout the wafer. From a processing point of view, equal-size contact windows will all tend to open at the same time during the etching process.

9.2.2 Polysilicon-Gate Transistor Layout

Transistors fabricated using polysilicon-gate technology have a number of important advantages over those built using metal-gate processes. We will discover some of these advantages by looking at the layout and structure of the polysilicon-gate transistor.

The mask sequence for the basic polysilicon-gate process from Chapter 1 is (again without passivation layer) as follows:

1. Active region (thin oxide) mask   First mask
2. Polysilicon mask   Align to level 1
3. Contact window mask   Align to level 2
4. Metal mask   Align to level 3

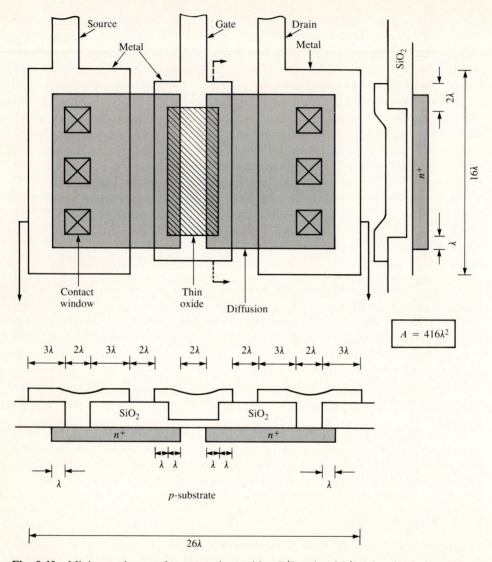

Fig. 9.10 Minimum-size metal-gate transistor with a Z/L ratio of $5/1$ using the design rules of Fig. 9.9. The active gate region is less than 5% of the total device area.

Some new design rules must be introduced for this process. Polysilicon lines and spaces will both be a minimum feature size of 2λ. The polysilicon gate must overlap the thin oxide region by an alignment tolerance λ. The above alignment sequence requires 1λ polysilicon and 1λ metal borders around contacts. However, contact holes should have a 2λ border of thin oxide due to tolerance accumulation.

Figure 9.11 shows the layout of the polysilicon-gate device with $Z/L = 5/1$ using these design rules. The total area is $168\lambda^2$. The active channel region now represents 12% of the total area, compared with less than 5% for the metal-gate device. The polysilicon gate acts as a barrier material during source/drain implantation and results in "self-alignment" of the edge of the gate to the edge of the source/drain regions. Self-alignment of the gate to the channel reduces the size of the transistor and eliminates the overlap region between the gate and the source/drain regions. In addition, the size of

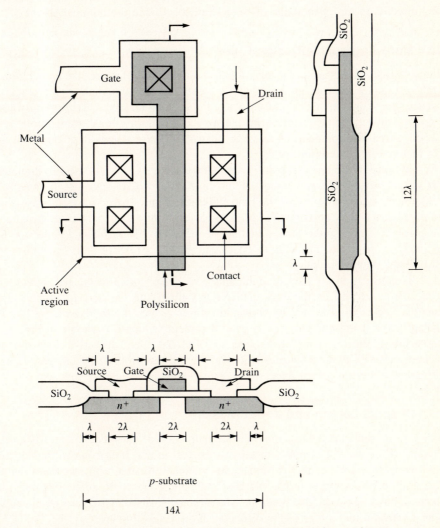

Fig. 9.11 Minimum-size polysilicon-gate transistor layout for $Z/L = 5/1$. The active gate region occupies 12% of the transistor area, and parasitic gate capacitance is minimized.

the transistor is reduced because the source/drain metallization can be placed nearer to the gate. In the polysilicon-gate layout, the channel is defined by the polysilicon gate in the length direction and by the thin oxide in the width direction.

A very important side benefit resulting from this process is the third level of interconnection provided by the polysilicon. Circuit wiring may be accomplished on the diffusion, metal, and polysilicon levels in the polysilicon-gate technology.

A design rule concerning edges has again been introduced into this layout. Metal lines are spaced 1λ from the polysilicon gate to prevent the edge of the metal line from falling directly on top of the edge of the polysilicon line in the nominal layout.

9.2.3 More-Aggressive Design Rules

The design rules discussed so far have focused on minimum feature size and alignment tolerance. F and T are determined primarily by the type of lithography being practiced. However, linewidth expansion and shrinkage throughout the process also strongly affect the ground rules. Expansion or shrinkage may occur during mask fabrication, resist exposure, resist development, etching, or diffusion. These linewidth changes are normally factored into the design rules.

In addition, alignment variation is a statistical process. Worst-case misalignments occur only a very small percentage of the time. (For a Gaussian distribution, a 3σ misalignment occurs only 2% of the time.) Our set of rules based on worst-case alignment tolerances is very pessimistic. For example, assuming that contacts are misaligned by λ in one direction, at the same time that the metal level is misaligned in the opposite direction by λ, results in an accumulated tolerance of 2λ. However, this situation would most probably never occur.

Let us consider the impact of tightening two design rules in the polysilicon-gate process. First, we will let the edge of one layer align with the edge of another layer. Second, a contact window will be allowed to run over onto the field oxide by 1λ. The resulting layout using our polysilicon-gate alignment sequence is shown in Fig. 9.12. The total area of the device has been reduced 25% to $120\lambda^2$, and the active channel region now represents 17% of the total transistor area. We see how ground rule changes can have a substantial effect on device area.

9.2.4 Channel Length and Width Biases

Figure 9.13 presents another example of the interaction of the process with design-rule definitions. Here we will assume a metal-gate process in which the source/drain junction depth is equal to λ and lateral diffusion equals vertical diffusion. Since we know that the source/drain diffusions will move laterally under the edge of the oxide openings, the contact windows can be aligned with the edge of the diffusions at the mask level but will still be 1λ within the border of the diffusion in the final structure.

However, lateral diffusion requires the length of channel at the mask level to be doubled to achieve the same electrical channel length in the device. The actual channel

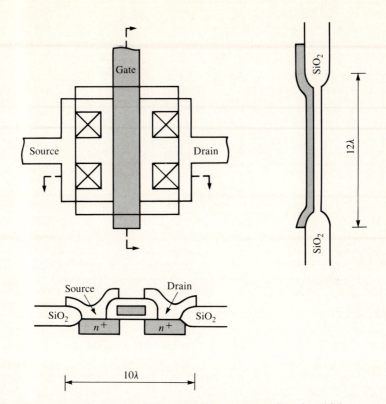

Fig. 9.12   More aggressive layout of the polysilicon-gate transistor in which two ground rules have been relaxed. Active gate area is now 17% of total device area.

length $L = L_m - \Delta L$, where L_m is the channel length as originally drawn on the mask and ΔL is the channel-length shrinkage which occurs during processing. This is an important area where the process must be controlled. For devices with short channel lengths, ΔL may be so severe that the devices become unusable. For the metal-gate layout of Fig. 9.13, $Z_m/L_m = 4\lambda/4\lambda = 1/1$ at the mask level, and $Z/L = 4\lambda/2\lambda = 2/1$ in the fabricated transistor.

The development of self-aligned polysilicon-gate technology with ion-implanted source/drain regions was a major improvement. The polysilicon-gate process eliminates most, but not all, of both the channel shrinkage caused by lateral diffusion and the overlap capacitance resulting from alignment tolerances in the metal-gate process.

In Fig. 9.11, one can see another source of channel bias. The "bird's beak" reduces the size of the active region to below that defined by the active region mask, and it introduces a process bias into the channel width of the polysilicon-gate transistor. $Z = Z_m - \Delta Z$, where Z_m is the width at the mask level and ΔZ is the channel-width shrinkage during processing.

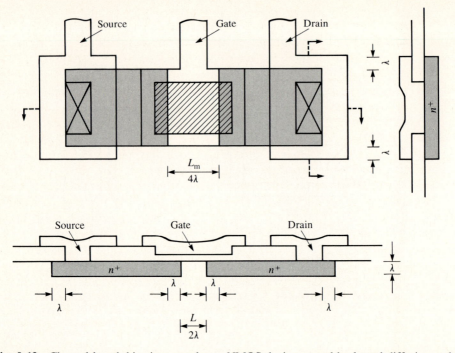

Fig. 9.13 Channel-length bias in a metal-gate NMOS device caused by lateral diffusion under the edge of the diffusion window. The transistor has $Z/L = 1/1$ at the mask level but ends up with an actual $Z/L = 2/1$ after the device is fabricated. Layout of the contact position is based on knowledge of the lateral diffusion which occurs during processing.

In sets of very tight design rules developed for high-volume-production ICs, all critical dimensions are adjusted to account for the processing and alignment sequences. This often results in a layout which must conform to a set of 50 to 100 design rules.[7] Such a set of design rules is highly technology-specific and cannot be transferred from one generation of lithography to the next. The Mead-and-Conway-style rules[7] reach a compromise between a set of rules which is overly pessimistic and wastes a lot of silicon area, and one that is extremely complex but squeezes out all excess area. The Mead-and-Conway-style design rules are being used for low-volume ICs in which design time, and not silicon area, is of dominant importance.

9.3 COMPLEMENTARY MOS (CMOS) TECHNOLOGY

The basic CMOS process of Fig. 1.5 requires a p-well diffusion and formation of both NMOS and PMOS transistors. Substrate resistivity is chosen to give the desired PMOS characteristics, and an additional implant step may be introduced to adjust the PMOS threshold separately. The *p*-well-to-substrate junction may range from a few microns to as much as twenty microns in depth. The net surface concentration of the *p* well must be high enough above the substrate concentration to provide adequate process control without severely degrading the mobility and threshold voltage of the NMOS transistors. The surface concentration of the *p*-well typically ranges between three and ten times the substrate impurity concentration. An additional implant step is often introduced to adjust the NMOS threshold voltage.

Parasitic bipolar devices are formed in the CMOS process in which merged *pnp* and *npn* transistors form a four-layer (*pnpn*) lateral SCR, as shown in Fig. 9.14. If this SCR is turned on, the device may destroy itself via a condition called *latchup*.[8, 9] The *p*-well depth and the spacings between the source/drain regions and the edge of the *p*-well must be carefully chosen to minimize the current gain of the bipolar transistors and the size of the shunting resistors R_s and R_w. A CMOS process will have a number of additional ground rules which are not present in an NMOS or PMOS process. A more detailed discussion of the design of bipolar transistors will be given in Chapter 10.

In order to reduce the resistance of the two shunting resistors, "guard ring" diffusions are sometimes added to the process, as in Fig. 9.14. Guard rings can be formed using the source/drain diffusions of the PMOS and NMOS transistors or can be added as

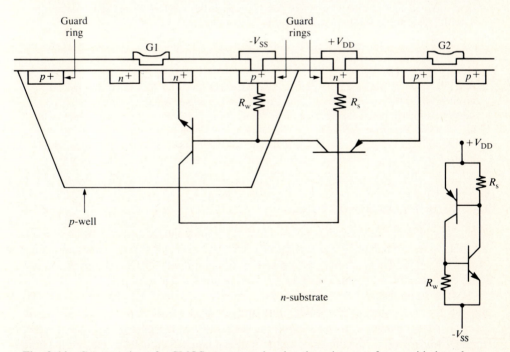

Fig. 9.14 Cross-section of a CMOS structure, showing the existence of a parasitic lateral *pnpn* SCR and the use of guard rings to reduce the value of R_s and R_w.

separate diffusion steps. Recent CMOS processes have used an *n*-well version of this technology, which permits optimization of the NMOS devices fabricated in the original substrate.

Twin-well processes, such as in Fig. 9.15, permit separate optimization of both NMOS and PMOS devices.[10] A lightly doped *n*- or *p*-type epitaxial layer is grown on a heavily doped *n*- or *p*-type substrate. (Lightly doped *n*- and *p*-type regions are often referred to as ν and π regions, respectively.) Separate implantations and diffusions are used to form wells for both the NMOS and PMOS transistors. The low-resistivity substrate substantially reduces the substrate resistance R_s and improves latchup resistance.

Example 9.2: A CMOS process uses an *n*-type substrate with a doping of $10^{15}/cm^3$. An implant/drive-in schedule will be used to form a *p*-well with a net surface concentration of $4 \times 10^{15}/cm^3$ and a junction depth of 7.5 μm. **(a)** What is the drive-in time at 1150 °C? **(b)** Solve for the implanted dose in silicon. **(c)** What are the threshold voltages of the *n*- and *p*-channel transistors if the oxide thickness is 50 nm?

Solution: The 7.5-μm junction depth and low surface concentration suggest that the well has a Gaussian profile resulting from a two-step diffusion or implant/diffusion process. A final surface concentration of $5 \times 10^{15}/cm^3$ is required to produce a net concentration of $4 \times 10^{15}/cm^3$ at the surface. Solving for the Dt product yields

$$Dt = x_j^2/2 \ln(N_0/N_B) = 8.74 \times 10^{-8} \ cm^2$$

At 1150 °C, $D = 8.87 \times 10^{-13} \ cm^2/sec$, which gives $t = 27.5$ h. The dose in silicon is given by $Q = N_0\sqrt{\pi Dt} = 2.62 \times 10^{12}/cm^2$. The *p*-channel devices reside in the *n*-type substrate with a doping concentration of $10^{15}/cm^3$. From Fig. 9.2, the threshold voltage will be -0.95 V. The deep well diffusion will be almost constant near the surface with a value of $4 \times 10^{15}/cm^3$. Figure 9.2 yields an *n*-channel threshold of 0.2 V. A threshold adjustment implant would be needed in this process to increase the *n*-channel threshold voltage.

9.4 OTHER MOS STRUCTURES

Chemical etching techniques may be used to crystallographically etch silicon. A solution of KOH, water, and alcohol[11] etches the $\langle 100 \rangle$, $\langle 110 \rangle$, and $\langle 111 \rangle$ crystal planes at relative rates of $40:30:1$. This etch may be masked by silicon dioxide or silicon nitride and can be used to etch cavities and V-shaped grooves in $\langle 100 \rangle$ silicon (Fig. 9.16).

VMOS technology[12] makes use of the grooves to reduce the channel length and increase the Z/L ratio of the MOS transistor. A basic VMOS process is shown in Fig. 9.17. The channel is formed along the four sides of the groove, and the channel length is determined by the thickness of the epitaxial layer and the diffusions. At the time VMOS was invented, channel lengths achievable with this technology were much shorter than those that could be achieved with normal planar technology, because the epitaxial

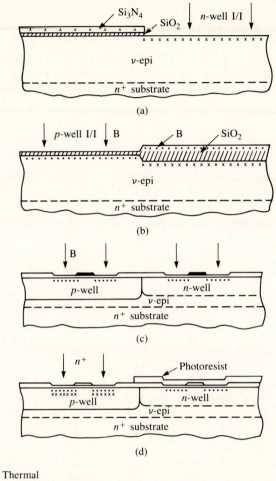

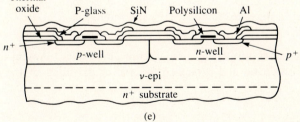

Fig. 9.15 Twin-well CMOS structure at several stages of the process. (a) n-well ion implant; (b) p-well implant; (c) nonselective p^+ source/drain implant; (d) selective n^+ source/drain implant using photoresist mask; (e) final structure. Copyright 1980 IEEE. Reprinted with permission from ref. [10].

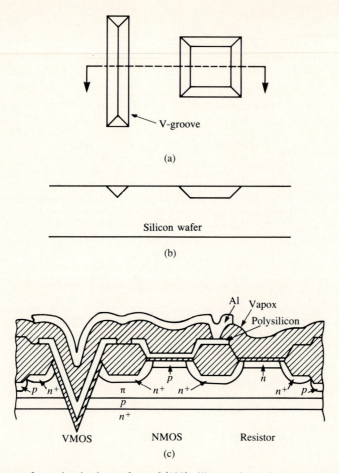

Fig. 9.16 Groove formation in the surface of ⟨100⟩ silicon using anisotropic etching of silicon. (a) Top view; (b) cross section; (c) use of a groove in the formation of a VMOS transistor. Copyright 1977 IEEE. Reprinted from ref. [12] with permission.

layer thickness was not limited by lithographic dimensions. Present-day MOS power transistors have improved and expanded the use of the ideas of the original VMOS structure.[13]

9.5 SUMMARY

In this chapter we have explored the interaction of process design with MOS device characteristics and transistor layout, including the relationships between processing parameters and breakdown voltage, punch-through voltage, threshold voltage, and junction capacitance. A low value of substrate doping is desired to minimize junction

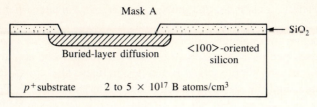

Mask A

Buried-layer diffusion $<100>$-oriented silicon

← SiO$_2$

p^+ substrate   2 to 5 $\times$ 10^{17} B atoms/cm^3

(a)

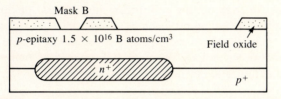

Mask B

p-epitaxy 1.5 $\times$ 10^{16} B atoms/cm^3 Field oxide

n^+

p^+

(b)

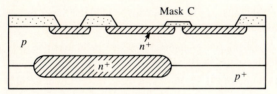

Mask C

p n^+

n^+

p^+

(c)

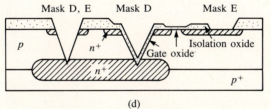

Mask D, E   Mask D   Mask E

p n^+ Isolation oxide

Gate oxide

n^+

p^+

(d)

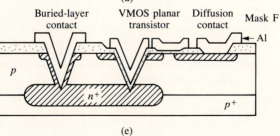

Buried-layer contact   VMOS planar transistor   Diffusion contact   Mask F

← Al

p

n^+

p^+

(e)

Fig. 9.17 Fabrication sequence for the formation of VMOS transistors for use in memory circuits. (a) Buried-layer diffusion; (b) epitaxial growth; (c) source/drain diffusion; (d) anisotropic etch and oxide growth; (e) metallization and pattern definition. Copyright 1978 IEEE. Reprinted with permission from ref. [17].

capacitance, substrate sensitivity, and junction breakdown voltage, whereas a high substrate doping is needed to maximize punch-through voltage. The use of ion implantation permits the designer to separately tailor the threshold voltage of the transistor.

We have developed basic ideas relating minimum feature size and alignment tolerances and have discussed simple sets of layout design rules. The strong relation between layout design rules and the size of transistors has been demonstrated. Polysilicon-gate technology has been shown to result in a much smaller device area than metal-gate technology for a given transistor Z/L ratio as well as to minimize the parasitic gate capacitance of the device. In addition, the polysilicon-gate process substantially reduces channel-length bias caused by lateral diffusion.

A combination of ion implantation and diffusion is commonly used to form the p- or n-well required for CMOS technology. VLSI CMOS often uses twin-well processes which permit separate optimization of both the n- and p-channel devices.

REFERENCES

[1] R. F. Pierret, *Field Effect Devices,* Volume IV in the Modular Series on Solid State Devices, Addison-Wesley, Reading, MA, 1983.

[2] S. A Abbas and R. C. Dockerty, "N-channel Design Limitations due to Hot Electron Trapping," IEEE IEDM Digest, p. 35–38 (1975).

[3] T. H. Ning, C. M. Osburn, and H. N. Yu, "Threshold Instability in IGFETs due to Emission of Leakage Electrons from Silicon Substrate into Silicon Dioxide," Applied Physics Letters, *29,* 198–199 (1976).

[4] P. E. Cottrell and E. M. Buturla, "Steady State Analysis of Field Effect Transistors via the Finite Element Method," IEEE IEDM Digest, p. 51–54 (1975).

[5] S. M. Sze, *Semiconductor Devices — Physics and Technology,* John Wiley & Sons, New York, 1985.

[6] C. A. Mead and L. Conway, *VLSI Design,* Addison-Wesley, Reading, MA, 1980.

[7] Brian Spinks, *Introduction to Integrated Circuit Layout,* Chapter 7, Prentice-Hall, Englewood Cliffs, NJ, 1985.

[8] A. Ochoa, W. Dawes, and D. Estreich, "Latchup Control in CMOS Integrated Circuits," IEEE Transactions on Nuclear Science, *NS-26,* 5065–5068 (December, 1979).

[9] R. S. Payne, W. N. Grant, and W. J. Bertram, "The Elimination of Latchup in Bulk CMOS," IEEE IEDM Digest, p. 248–251 (December, 1980).

[10] L. C. Parrillo, R. S. Payne, R. E. Davis, G. W. Reutlinger, and R. L. Field, "Twin-Tub CMOS — A Technology for VLSI Circuits," IEEE IEDM Digest, p. 752–755 (December, 1980).

[11] E. Bassous, "Fabrication of Novel Three-Dimensional Microstructures by the Anisotropic Etching of ⟨100⟩ and ⟨110⟩ Silicon," IEEE Transactions on Electron Devices, *ED-25,* 1178–1185 (October, 1978).

[12] T. J. Rodgers, F. B. Jenne, B. Frederick, J. J. Barnes, W. R. Hiltpold, and J. D. Trotter, "VMOS Memory Technology," IEEE ISSCC Digest, p. 74–75 (February, 1977).

[13] B. J. Baliga and D. Y. Chen, *Power Transistors: Device Design and Applications,* IEEE Press, New York, 1984.

[14] K. P. Roenker and L. W. Linholm, "An NMOS Test Chip for a Course in Semiconductor Parameter Measurements," National Bureau of Standards Internal Report 84–2822 (April, 1984).

[15] T. J. Russell, T. F. Leedy, and R. L. Mattis, "A Comparison of Electrical and Visual Alignment Test Structures for Evaluating Photomask Alignment in Integrated Circuit Manufacturing," IEEE IEDM Digest, p. 7A–7F (December, 1977).

[16] D. S. Perloff, "A Four-Point Electrical Measurement Technique for Characterizing Mask Superposition Errors on Semiconductor Wafers," IEEE Journal of Solid-State Circuits, *SC–13*, 436–444 (August, 1978).

[17] K. Hoffman and R. Losehand, "VMOS Technology Applied to Dynamic RAMs," IEEE Journal of Solid-State Circuits, *SC–13*, 617–622 (October, 1978).

PROBLEMS

9.1   What is the maximum gate-to-source voltage that a MOSFET with a 10-nm gate oxide can withstand. Assume that the oxide breaks down at 5 MV/cm and that the substrate voltage is zero.

9.2   Two n^+ diffused lines are running parallel in a substrate doped with 10^{15} boron atoms/cm³. The substrate is biased to -5 V, and both lines are connected to $+5$ V. Using one-dimensional junction theory, calculate the minimum spacing needed between the lines to prevent their depletion regions from merging.

9.3   Use one-dimensional junction theory to estimate the punch-through voltage of a MOSFET with a channel length of 1 μm. Assume a substrate doping of 3×10^{16}/cm³ and a substrate bias of 0 V.

9.4   Calculate the threshold voltage for the NMOS transistor with the doping profile shown in Fig. P9.4. Assume an n^+ polysilicon-gate transistor with a gate-oxide thickness of 50 nm.

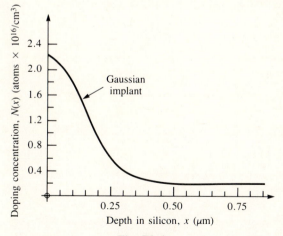

Fig. P9.4

9.5   An *n*-well CMOS process starts with a substrate doping of $3 \times 10^{15}/\text{cm}^3$. The well doping near the surface is approximately constant at a level of $3 \times 10^{16}/\text{cm}^3$. The gate-oxide thicknesses are both 40 nm.

(a) Calculate the thresholds of the *n*- and *p*-channel transistors using eqs. (9.2).

(b) Calculate the boron doses needed to shift the NMOS threshold to $+1$ V and the PMOS threshold to -1 V. Assume that the threshold shifts are achieved through shallow ion implantations. Neglect oxide charge.

9.6   High-performance NMOS logic processes use depletion-mode NMOS transistors for load devices. This requires a negative threshold which can be obtained by implanting a shallow arsenic or phosphorus dose into the channel region. Calculate the arsenic dose needed to achieve a -3-V threshold in an n^+ polysilicon-gate NMOS transistor which has a substrate doping of $3 \times 10^{16}/\text{cm}^3$ and a gate-oxide thickness of 50 nm.

9.7   Draw a composite view of the situation resulting from a worst-case misalignment of the masks for the MOSFET layout shown in Fig. 9.10. Assume metal aligns to thin oxide, and thin oxide and contacts align to the diffusion.

9.8   Develop a new set of ground rules for the metal-gate transistor of Section 9.2, assuming that levels 2, 3, and 4 are all aligned to level 1. Redraw the transistor of Fig. 9.10 using your new rules. In what ways is this layout better or worse than that originally given in Fig. 9.10?

9.9   Draw a cross section of a metal-gate NMOS transistor and a composite view of its mask set, assuming an aggressive layout which takes into account all lateral diffusion. Assume a source/drain junction depth of 2.5 μm and assume that lateral diffusion equals 80% of vertical diffusion. Assume λ is 2 μm and $Z/L = 10/1$.

9.10   Draw the layout of a three-input NMOS NOR-gate with the dimensions given on the circuit schematic in Fig. P9.10. Be sure to merge diffusions wherever possible. Use the more aggressive ground rules developed for polysilicon-gate devices.

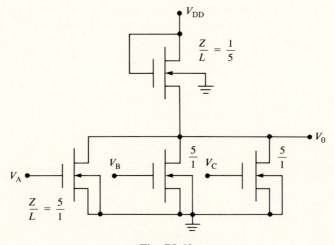

Fig. P9.10

9.11   Our design rule examples used an alignment tolerance which was one-half the feature size. This ratio represents a very loose alignment capability. Develop a new set of design rules similar to those of Fig. 9.11 for $T = \alpha$ and $F = 4\alpha$. Draw the new minimum-size polysilicon-gate transistor using your rules. Compare the area of your transistor with the area of the transistor of Fig. 9.11 if $\lambda = 2\alpha$.

9.12 An implant with its peak concentration at the silicon surface is used to adjust the threshold of an NMOS transistor. We desire to model this implant by a rectangular approximation similar to that of Figure 9.5. Show that $N_i = N_p \pi/4$ and $x_i = \Delta R_p \sqrt{8/\pi}$ by matching the first two moments of the two impurity distributions.

9.13 A number of types of alignment test structures have been developed.[14, 15] Figure P9.13 shows a simple test structure which can be used to measure the misregistration of the contact

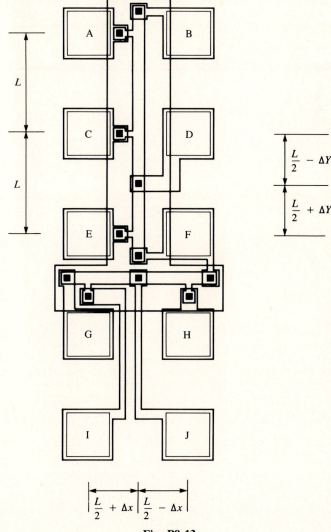

Fig. P9.13

window mask relative to the diffusion mask.[16] Two linear potentiometers, one in the horizontal direction and one in the vertical direction, are fabricated using diffused resistors. The distance between contacts A and C is the same as that between C and E, and the contact from pad D is nominally one-half the distance between pads C and E. A current is injected between pads B and F, and the voltages between pads C-D and D-E are measured.

(a) Show that the misregistration in the y direction is given by $\Delta Y = \frac{1}{2} L \ (V_{DE} - V_{CD})/V_{AC}$.

(b) Derive a similar relationship for misregistration in the x direction.

10 / Bipolar Process Integration

In this chapter, interactions between fabrication processes and bipolar device design and layout will be explored. In particular, we will look closely at relationships between impurity profiles and device parameters such as current gain, transit time, and breakdown voltage. The use of recessed oxidation and self-aligned processes in the formation of high-performance bipolar transistors will be presented, and design rules for bipolar structures will also be discussed. In conclusion, collector-diffusion, V-groove, and dielectric isolation processes will be discussed.

10.1 THE JUNCTION-ISOLATED STRUCTURE

The basic junction-isolated bipolar process of Fig. 10.1a has been used throughout the IC industry for many years and has become known as the *standard buried collector* (SBC) process. In this junction-isolated process, adjoining devices are separated by back-to-back *pn* junction diodes which must be reverse-biased to ensure isolation (see Fig. 10.1b). The SBC process remains the primary bipolar process for analog and power circuit applications with power supplies exceeding 15 V. Although the SBC process was also originally used for logic circuits, most digital technologies have evolved to self-aligned, oxide-isolated processes using polysilicon and other technology advances first developed for MOS processes.

The process flow for the SBC structure of Fig. 10.1a was discussed in Section 1.4 and will only be outlined here. An n^+ buried layer is formed by selective diffusion into a $\langle 111 \rangle$-oriented *p*-type substrate and is followed by growth of an *n*-type epitaxial layer. Isolated *n*-type collector islands are formed using a deep boron diffusion. The base and emitter are formed by successive *p*- and *n*-type diffusions into the epitaxial layer. The structure is completed with contact window formation and metallization.

A cross section of the SBC impurity profile through the center of the device is shown in Fig. 10.2. In the next several sections we will consider how the design of this profile is related to several important measures of device performance. An understanding of the basic profile design for the SBC process will help us see the advantages and disadvantages of other types of processes.

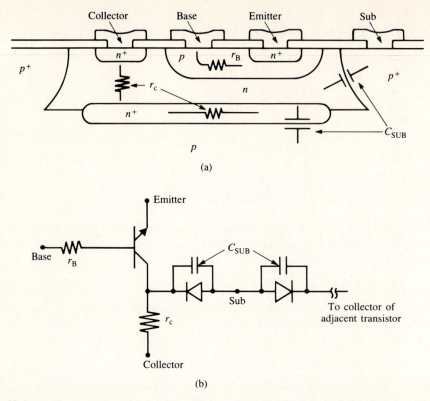

Fig. 10.1   (a) Cross section of a transistor fabricated with the SBC process showing the collector-base capacitances and the base and collector series resistances; (b) lumped circuit model for the transistor showing back-to-back diodes which provide isolation between adjacent transistors.

10.2   CURRENT GAIN

In order to be useful in circuits, the bipolar transistor must have a current gain of at least 10 to 20 for digital applications and an order of magnitude greater for analog applications. An expression for the current gain of the bipolar transistor is

$$\beta^{-1} = \frac{G_B}{G_E} + \frac{W_B^2}{\eta L_B^2} \tag{10.1}$$

The basewidth, W_B, is the width of the electrically neutral base region of the transistor. The constant η is determined by the shape of the impurity profile in the base and ranges from 2 to 20. L_B is the diffusion length for minority carriers in the base. G_B and G_E are called the *Gummel numbers* in the base and emitter respectively.

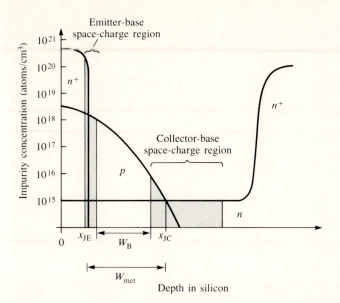

Fig. 10.2 Vertical impurity profile in typical bipolar junction transistor. The shaded regions represent the emitter-base and collector-base space-charge regions. The metallurgical basewidth and electrical basewidth are indicated by W_{met} and W_B, respectively.

The *Gummel numbers* are defined by

$$G_B = \int_{base} \frac{N(x)}{D_B(x)} dx \quad \text{and} \quad G_E = \int_{emitter} \frac{N(x)}{D_E(x)} dx \qquad (10.2)$$

where D_E and D_B are the minority-carrier diffusion constants in the emitter and base. Heavy doping effects in the emitter typically limit the value of G_E to 10^{13} to 10^{14} sec/cm^4. The basewidth is defined by the distance between the edges of the two space-charge regions in the base. For wide-base transistors, this is approximately equal to the distance between the metallurgical junctions as shown in Fig. 10.2. For narrow-base transistors, the space-charge regions must be subtracted from the metallurgical basewidth, as discussed further in Section 10.4.

For large current gain, eq. (10.1) should be as small as possible. The ratio of the Gummel numbers in the base and emitter should be low, the width of the base region should be small, and L_B should be large. Figure 10.3 shows the dependence of the diffusion length on impurity concentration. As the doping level increases, L_B decreases, but is greater than 10 μm for typical base-doping concentrations. In modern high-frequency transistors, the basewidth W_B is typically much less than the diffusion length L_B, and the first term in eq. (10.1) determines the current gain.

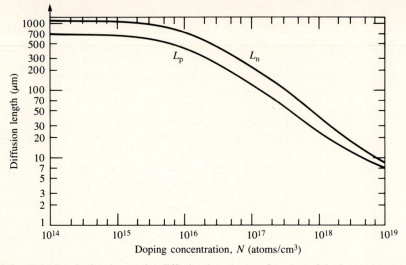

Fig. 10.3 Calculated minority-carrier diffusion lengths as a function of doping concentration for bulk silicon using lifetime equations from ref. [1].

From eqs. (10.1) and (10.2), the emitter must be heavily doped relative to the base in order to obtain high gain. In fabricating a bipolar transistor, each successive diffusion is heavier than the last, and the final n^+ diffusion naturally performs best as the emitter.

Example 10.1: Estimate the current gain for a transistor with the following parameters: $N_E/D_E = 5 \times 10^{13}$ sec/cm^4, $N_B/D_B = 10^{12}$ sec/cm^4, $W_B = 1$ μm, $L_B = 20$ μm, and $\eta = 10$.

Solution: Plugging these parameters into eq. (10.1) yields $\beta^{-1} = 0.02 + 0.00025$ and $\beta = 50$. In this transistor, the current gain is dominated by the ratio of the Gummel number terms.

10.3 TRANSIT TIME

Another important bipolar device parameter is the delay incurred during carrier propagation between the emitter and collector terminals of the transistor. Both logic switching speed and amplifier frequency response are limited by the *transit time,* which is defined by

$$\tau = W_B^2/\eta D_B + (C_{JC} + C_{sub})r_C + X_C/2V_S \tag{10.3}$$

The unity-gain frequency of the transistor, f_T, is given approximately by

$$f_T = 1/2\pi\tau \tag{10.4}$$

The first term in eq. (10.3), called the *base transit time,* represents the time required for a carrier to move across the neutral base region W_B. The second term is the delay associated with charging the capacitances connected to the collector node through the collector series resistance r_C. The capacitances C_{JC} and C_{sub} are determined by the collector-base and collector-substrate junction areas and by the doping concentrations of the base, collector, and substrate regions. The third term is the delay time associated with a carrier crossing the depletion region of the collector-base junction. X_C is the width of the depletion layer and V_S is the saturation velocity of the carriers.

In order to minimize τ, the basewidth is made as narrow as possible, the buried layer is added to minimize the value of r_C, and light doping is used to minimize the capacitances. Estimates of the capacitance of the junctions can be made using the one-sided step-junction expression (eq. 9.3), in which the capacitance is determined by the concentration on the lightly doped side of the junction.

Example 10.2: Calculate the transit time for a bipolar transistor with the following parameters: $W_B = 1 \ \mu m$, $\eta = 10$, $D_B = 20 \ cm^2/sec$, $C_{JC} + C_{sub} = 2 \ pF$, $r_C = 250$ ohms, $X_C = 10 \ \mu m$, and $V_S = 10^7 \ cm/sec$.

Solution: Plugging these values into eq. (10.3) gives the following values for the three terms: (i) 0.05×10^{-9} sec; (ii) 0.5×10^{-9} sec; (iii) 0.05×10^{-9} sec. The resulting value of transit time is 0.60×10^{-9} sec. The unity-gain frequency f_T is equal to 265 MHz.

10.4 BASEWIDTH

Eqs. (10.1) through (10.3) indicate that device performance is improved by making the basewidth as narrow as possible. The primary restrictions on reducing the basewidth are set by breakdown-voltage requirements and by tolerances on the basewidth due to variations in process control. For low-voltage logic devices, the metallurgical basewidth may be less than 1 μm. For higher-voltage devices used in analog circuit or power applications, the basewidth must be wide enough to support the collector-base depletion-layer width under large reverse bias.

The actual basewidth of the transistor is determined by reducing the metallurgical basewidth by the portions of the emitter-base and collector-base space-charge regions which protrude into the base, as shown in Fig. 10.2. The emitter and base are both heavily doped near the junction, and although the space-charge-region width of the emitter-base junction is usually quite small, it does extend almost entirely into the base. Its width can be estimated from Fig. 9.4.

The collector-base space-charge region width is dependent on the voltage across the junction and extends into both the base and collector regions. Figure 10.4 shows the depletion-layer width on either side of a *pn* junction formed by a Gaussian diffusion into a uniformly doped substrate, the normal situation for a bipolar transistor fabricated using the SBC process.

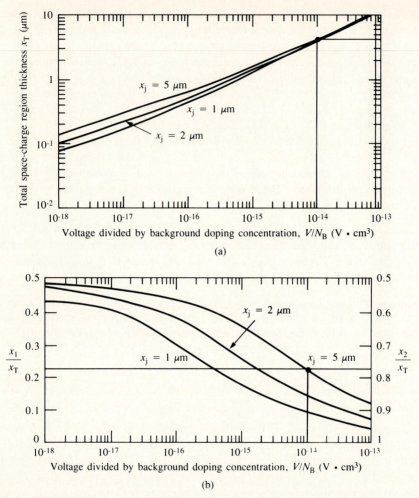

Fig. 10.4 The space-charge region width as a function of voltage and doping for a *pn* junction formed by a Gaussian diffusion into a uniformly doped substrate. (a) Total space-charge region width x_T; (b) fraction of total space-charge region width extending on the heavily doped side, x_1, and on the lightly doped side, x_2, respectively. After ref. [4]. Reprinted with permission from the *AT&T Technical Journal*. Copyright 1960 AT&T.

Example 10.3: Estimate the space-charge region widths on each side of the collector-base junction of a bipolar transistor fabricated on a 1-ohm-cm *n*-type epitaxial layer. The reverse-bias voltage across the junction is 40 V, and the collector-base junction depth is 5 μm.

Solution: The doping of the epitaxial layer is $4 \times 10^{15}/\text{cm}^3$, giving a value of $V/N_B = 1 \times 10^{-14}$ V-cm^3. From Fig. 10.4a, the total depletion-layer width is approximately 4 μm. From Fig. 10.4b, 77% or 3.1 μm extends into the collector region, and 23% or 0.9 μm extends into the base region.

Heavy base doping reduces the size of the space-charge regions in the base, permitting a narrow-base design. However, heavy base doping tends to increase the Gummel number in the base, which reduces the current gain of the transistor. Heavy doping also increases the collector-junction capacitance, thus increasing the transit time in eq. (10.3). This is another situation in which conflicts arise when trying to optimize several different device parameters simultaneously.

10.5   BREAKDOWN VOLTAGES

The process designer must understand the magnitude of the voltages that will be applied to the transistors in circuit applications. The device in Example 10.3 had to withstand 40 V and was probably designed for analog-circuit applications. On the other hand, transistors designed for logic applications must support only relatively low voltages. For example, the devices used in TTL circuits are designed to withstand only 7 V, and devices for some ECL gates can break down at even lower voltages.

10.5.1   Emitter-Base Breakdown Voltage

Emitter-base breakdown voltage is determined by the doping concentration and radius of curvature of the junction, as was discussed previously in Section 9.1.2. Breakdown occurs first in the region of the junction where the electric field is the largest, usually corresponding to the portion of the junction where the doping levels are the highest. The actual breakdown voltage is then determined by the doping on the more lightly doped side of the junction.

In order to achieve high current gain, the emitter region is doped heavily, and the breakdown voltage of this junction will be determined by the impurity concentration of the more lightly doped base region. The base impurity concentration is highest at the surface, so the emitter-base junction will tend to break down first at the surface. The curvature of the junction enhances the electric field and reduces the breakdown voltage. Figure 10.5 gives the breakdown voltage of the emitter-base junction as a function of the final surface concentration of the base region, with junction radius as a parameter. Emitter-base breakdown voltages are low because of the relatively large impurity concentrations on both sides of the junction.

Example 10.4: An *npn* transistor has a 1-μm-deep emitter-base junction and the base diffusion given in Example 4.2. What is the expected breakdown voltage of this junction?

Solution: The base-region surface concentration in Fig. 4.9 is 1.1×10^{18}/cm^3. Figure 10.5 predicts the breakdown voltage of a 1-μm-deep junction with this surface

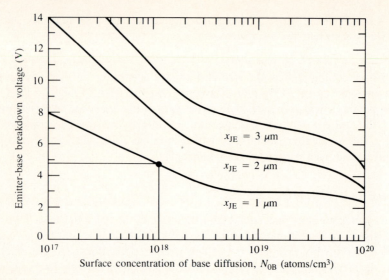

Fig. 10.5   Emitter-base junction breakdown voltage as a function of base surface concentration with emitter-base junction depth as a parameter. After ref. [4]. Reprinted with permission from *Solid-State Electronics,* Vol. 17, P. R. Wilson, "The Emitter-Base Breakdown Voltage of Planar Transistors," Copyright 1974, Pergamon Press, Ltd.[3]

concentration to be approximately 4.8 V. (The emitter-base junction of most common bipolar transistors will break down well below 10 V.)

10.5.2   Collector-Base Breakdown Voltage

The bipolar transistor can begin to conduct excessive collector current by two mechanisms. The first is Zener or avalanche breakdown of the collector-base junction. As discussed above, breakdown is localized to the region where the doping concentrations are the largest, but it is determined primarily by the doping concentration on the more lightly doped side of the junction. The collector is formed in the uniformly doped epitaxial layer. The base region is diffused into the epitaxial layer and is the more heavily doped side of the junction. Since the collector is uniform, junction breakdown will occur first where the electric field is enhanced by junction curvature. The breakdown voltage for the collector-base junction as a function of epitaxial-layer impurity concentration and junction radius is given in Fig. 10.6.

The second breakdown mechanism is punch-through of the base region, illustrated in Fig. 10.7. The epitaxial layer is more lightly doped than the base region, and the collector-base junction depletion layer extends predominantly into the epitaxial layer. As the collector-base voltage increases, the depletion layer expands further into the epitaxial

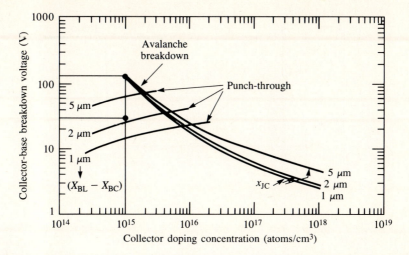

Fig. 10.6 Collector-base junction breakdown voltage as a function of collector-doping concentration with collector-base junction depth and punch-through limits as parameters. After ref. [4]. Reprinted from the JOURNAL OF THE ELECTROCHEMICAL SOCIETY, Volume 113 (1966), pages 508–510, by permission of the publisher, The Electrochemical Society, Inc.,[11] and Pergamon Press, Ltd.[12]

layer and will eventually hit the n^+ buried layer. At this point, further depletion-layer expansion will occur in the base, and any increase in collector-base voltage will quickly punch through the remaining base region.

The second set of curves in Fig. 10.6 shows collector-base junction breakdown limitations set by punch-through. As described above, the primary parameters determining the punch-through voltage are the epitaxial-layer doping and the width, $X_{BL} - X_{BC}$, of the region between the collector-base junction and the n^+ buried layer.

Example 10.5: What is the collector-base breakdown voltage of a transistor with a 10-μm-thick epitaxial layer doped at a level of $10^{15}/cm^3$ if the collector-base junction depth is 5 μm? Assume that the buried layer has diffused upward 2 μm.

Solution: First, determine the avalanche breakdown voltage of an isolated *pn* junction. Figure 10.6 gives a breakdown voltage of approximately 130 V for a doping of $10^{15}/cm^3$. Next, we must also check the punch-through limitations for $X_{BL} - X_{BC} = 3$ μm. From Fig. 10.6, the transistor will punch through at approximately 30 V. So the collector-base breakdown voltage is limited to 30 V by punch-through in this transistor.

As usual, different device requirements produce conflicting design constraints. High Zener breakdown voltage requires low epitaxial-layer doping. Low epitaxial-layer doping requires a relatively wide epitaxial-layer thickness in order to prevent punch-through.

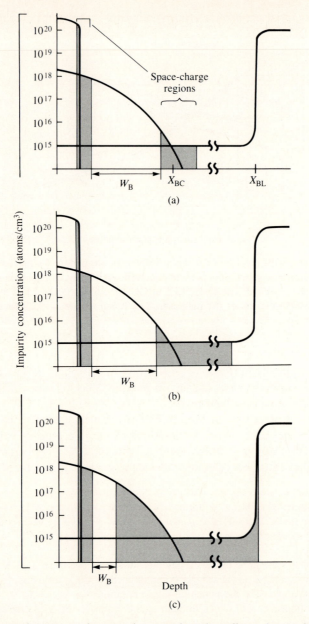

Fig. 10.7 Collector-base space-charge region growth as the collector-base voltage is increased. (a) Zero bias; (b) intermediate collector-base voltage; (c) large collector-base voltage just below the punch-through voltage.

However, a wide depletion-layer width in the epitaxial layer increases the transit time and reduces the frequency response of the device.

10.6   LAYOUT CONSIDERATIONS

This section will explore mask layout for the SBC transistor. The analysis will expand our understanding of the interaction of process design and layout. In particular, the top view of the mask set for a bipolar transistor often differs greatly from the final device structure due to large lateral diffusions, although this is less true of the high-performance digital technology to be discussed in Section 10.7.

10.6.1   Buried-Layer and Isolation Diffusions

The spacing between adjacent buried layers and the width of the intervening isolation diffusion determines how closely two transistors can be spaced. To maintain electrical isolation, the substrate is tied to the most negative voltage present in the circuit. The collector of a transistor, on the other hand, is often connected to the most positive voltage in the same circuit. Thus, the collector-substrate junction must be designed to support a voltage equal to the sum of the positive and negative voltages supplying the circuit. For analog circuits, this is typically more than 40 V. A typical design value of 60 V would provide an adequate safety margin.

Figure 10.8 shows the depletion layer in the p and n material near the isolation region of the transistor. The doping of the isolation diffusion is heavy at the surface and intersects the original substrate to produce isolation. At the surface, the window defining the isolation diffusion may be a minimum feature size, but the total width of the isolation region at the surface will be determined by lateral diffusion. For example, if the epitaxial layer is 15 μm thick and the minimum feature size is 10 μm, the isolation region will approach 40 μm in width, assuming lateral diffusion equals vertical diffusion.

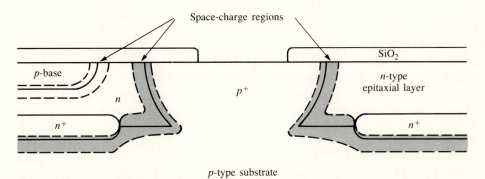

Fig. 10.8   Isolation region between two bipolar transistors. The spacing must be large enough to ensure that the two space-charge regions do not merge together.

The n^+ buried-layer diffusion is not usually permitted to intersect the p^+ isolation diffusion. If the two diffusions meet, the breakdown voltage of the junction will decrease, and the capacitance of the junction will increase. Thus, there will be a layout design rule associated with the minimum spacing between the n^+ region and the isolation diffusion.

10.6.2 Base Diffusion to Isolation Diffusion Spacing

At the surface, the collector-base and collector-substrate depletion regions of Fig. 10.8 must not merge. The minimum spacing can be determined from a knowledge of the applied voltages and the epitaxial-layer impurity concentration. Additional spacing must be added to account for the alignment sequence and accumulated alignment tolerances.

Example 10.6: What is the minimum spacing between the edge of the base diffusion and the edge of the isolation diffusion at the surface of a bipolar transistor if the alignment tolerance is 5 μm and the epitaxial-layer resistivity is 10 ohm-cm? Assume the two junctions must each support 40 V. Use a collector-base junction depth of 5 μm.

Solution: A 10-ohm-cm epitaxial layer has an impurity concentration of $5 \times 10^{14}/cm^3$, giving a value of $V/N_B = 8 \times 10^{-14}$ V-cm^3. From Fig. 10.6, the total depletion-layer width is approximately 10 μm, with 8.7 μm in the epitaxial layer. The conditions at the isolation-collector junction are essentially the same, so the minimum spacing will be two times the depletion-layer width of 8.7 μm plus the alignment tolerance of 5 μm for a total of 22.4 μm.

10.6.3 Emitter-Diffusion Design Rules

The minimum spacing between the edges of the emitter and base diffusions must be greater than the sum of the emitter and collector depletion-layer widths in the base, the accumulated alignment tolerance between the emitter and base masks, and the active base-region width.

In the basic SBC process, the n^+ emitter diffusion is also used to ensure formation of a good ohmic contact to the collector, and this collector contact diffusion should not intersect the depletion layers associated with either the p-type base or isolation diffusions. If this occurs, the breakdown voltage of the junctions will be reduced and the junction capacitances increased. (See Problem 10.7.)

10.6.4 A Layout Example

A set of design rules for a hypothetical bipolar transistor is given in Table 10.1, and Fig. 10.9 shows the layout of a minimum-size transistor based on these rules and making maximum use of lateral diffusion. It is interesting to note that the active area of the transistor, the region directly under the emitter, is a small fraction of the total device area

Table 10.1 Bipolar Transistor Design Rules for Fig. 10.9.

Minimum feature size	5 μm
Worst-case alignment tolerance between levels	2 μm
Epitaxial-layer thickness	10 μm
Collector-base junction depth	5 μm
Emitter-base junction depth	3 μm
Minimum emitter-to-collector spacing at surface	5 μm
Minimum base-to-isolation spacing at surface	5 μm
Minimum collector contact n^+ diffusion to isolation spacing	5 μm
Minimum collector contact n^+ diffusion to base spacing	5 μm
Buried-layer diffusion (both up and down)	2 μm
Buried layer to isolation spacing	5 μm
Lateral diffusion = vertical diffusion	

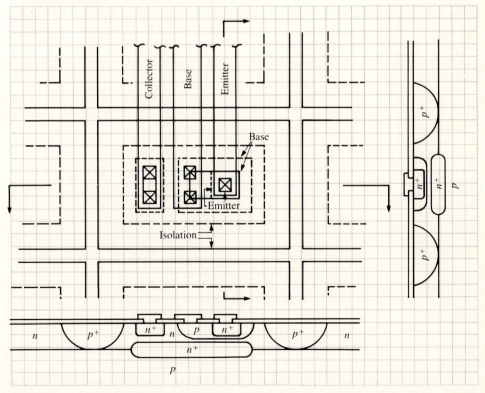

Fig. 10.9 Minimum-size bipolar transistor layout based on the design rules of Table 10.1. The buried layer is not shown in the top view for reasons of clarity. Each square is 5 μm x 5 μm.

of approximately 4536 μm^2. The final emitter area is 11 × 11 μm, or 121 μm^2, which represents only 2.67% of the total area of the transistor.

The rest of the area is needed to make contacts, to support depletion layers, and to provide isolation between adjacent devices. In this layout, the isolation area is 2800 μm^2, or more than 60% of the total area! Minimization of the isolation region represents an important issue in high-performance devices, not only for density improvement but also for junction-capacitance reduction.

The solid lines in the top view of the transistor layout represent the edges of the masks used to fabricate the transistor. The various dotted lines represent the final positions of the emitter, base, and isolation diffusions. For the design rules of Table 10.1, the window for the emitter diffusion happens to coincide exactly with the emitter contact window. The lateral diffusion of 3 μm provides more than the required 2-μm alignment tolerance for the emitter contact window.

The base diffusion is 5 μm deep and is assumed to diffuse laterally 5 μm. In the layout, the base contact windows actually extend outside the base region at the mask level but are more than one alignment tolerance within the base region following diffusion. This is an excellent example of the interaction between processing and layout.

The width of the base-contact metallization has been widened to more than a minimum feature size to help clarify the figure. This did not affect the size of the device, as space was available because of other design rule limitations. Two base and two collector contact windows fit within the minimum base and n^+ collector contact regions. The collector contact windows align with the edges of the n^+ diffusion window, as was the case for the emitter.

The buried-layer mask has also been omitted from the figure for clarity. In this particular structure, the design rules relating to the buried layer are not limiting factors in the size of this layout.

10.7 ADVANCED BIPOLAR STRUCTURES

For digital logic circuits, structures are optimized to provide as short a transit time as possible. This requires minimizing the basewidth, eliminating as much capacitance as possible by minimizing total junction area, minimizing the width of the collector space-charge region, and reducing the collector and base series resistances. A reduced current gain is traded for a shorter transit time.

Figure 10.10 shows a high-performance bipolar structure which attempts to achieve these goals by using a very thin epitaxial layer and shallow ion-implanted base and emitter regions. As much pn junction area as possible is eliminated through the use of oxide isolation. The sides of the emitter and base regions are actually walled by the oxide isolation regions. The n^+ buried layer is relatively large to minimize r_C, and the total base region is minimized to reduce the base resistance. Self-aligned contacts are made to the base, emitter, and collector regions.

The formation of the transistor of Fig. 10.10 begins with implantation and diffusion of the buried layer with a typical sheet resistance of 10 to 50 ohms per square. The

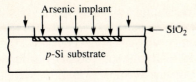

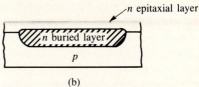

(a)   (b)

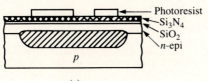

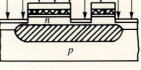

(c)   (d)

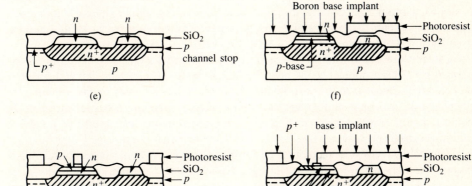

(e)   (f)

(g)   (h)

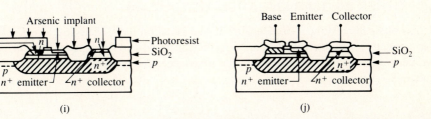

(i)   (j)

Fig. 10.10 Process sequence for a high-performance oxide-isolated bipolar transistor. (a) Buried-layer formation; (b) epitaxial layer growth; (c) mask for selective oxidation; (d) boron implant prior to recessed oxide growth; (e) selective oxidation; (f) base mask and boron base implantation; (g) emitter, base contact, and collector contact mask; (h) p^+ base contact implantation; (i) arsenic implantation for emitter and collector contact; (j) structure completed with multilayer metallization. Copyright, 1985, John Wiley & Sons, Inc. Reprinted with permission from ref. [5].

masking oxide is removed, and a thin epitaxial layer is grown on the surface. A recessed oxide isolation process is used to form the isolation regions between devices and to eliminate unnecessary junction area between the collector and emitter contacts. Prior to oxidation, part of the epitaxial layer is etched away so that the subsequent oxidation will extend completely through the epitaxial layer. An implantation is used to overcome boron depletion in the substrate during oxidation.

Next, the silicon nitride–oxide sandwich is removed, and an oxide is regrown on the surface. A boron implantation creates the shallow active base region. A mask is used to create windows for the emitter, and contacts to the base, emitter, and collector are all defined at the same time. Note that a single oxide strip defines both the emitter and base contact regions, eliminating alignment tolerances that would be needed if the regions were formed separately. The width of this strip is set by the metal-to-metal spacing plus accumulated alignment tolerances.

Photoresist is used as a barrier material during implantation of the base contact region. This p^+ implantation further reduces the base resistance of the device. Photoresist is also used as a barrier material during implantation of the emitter and collector contact regions. Note that these two mask steps use noncritical *blockout* masks similar to those used for threshold adjustment in a CMOS process.

Contacts are made through the same openings used for the base and emitter implantations. These contact areas are all cleared by a short wet or dry etch prior to metallization. Because of the very shallow junction depths involved in this structure, the metallization will be a multilayer sandwich structure including a barrier metal in the contact region. Interconnection of devices to form circuits will typically involve a multilevel metal process.

Figure 10.11 shows another high-performance structure which uses deep-trench isolation. The trenches are formed using reactive-ion etching and are refilled with polysilicon and silicon dioxide. In this structure, the emitter, and the contacts to the base and collector, are all formed by impurity diffusion from doped polysilicon.[6]

10.8   OTHER BIPOLAR ISOLATION TECHNIQUES

Several other interesting approaches to device isolation have been developed over the years, and three are surveyed here. The dielectric isolation process is still in use in high-performance analog circuits; the others have been largely replaced by oxide-isolated structures.

10.8.1   Collector-Diffusion Isolation (CDI)

The collector-diffusion-isolation[7] structure, shown in Fig. 10.12, was developed primarily for digital applications. The process eliminates the p-type isolation diffusion, achieving reduced device area and process complexity.

The process starts with diffusion of a low-sheet-resistance buried layer which will serve as the collector of the transistor. A thin p-type epitaxial layer forms the base region

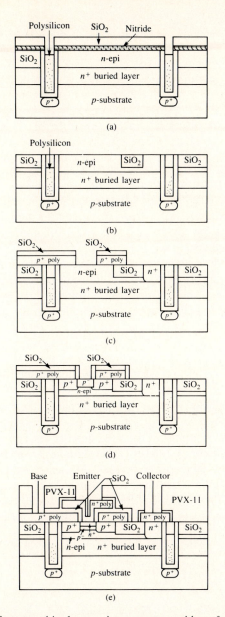

Fig. 10.11 A high-performance bipolar transistor structure with an f_T of 10 GHz. (a) Isolation is achieved using deep-trench isolation with polysilicon and silicon dioxide refill; (b) structure following selective oxidation; (c) p^+ polysilicon deposited and patterned; (d) diffusion from doped polysilicon forms the extrinsic base region and base contacts; a self-aligned implantation forms the intrinsic base; (e) diffusion from n^+ polysilicon forms the emitter and the emitter and collector contacts of the transistor. Copyright 1985 IEEE. Reprinted with permission from ref. [6].

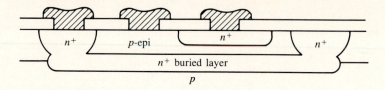

Fig. 10.12   Cross section of a transistor fabricated in the CDI process.[7]

and is grown in the next step. Device isolation is achieved via an n^+ diffusion which completely encloses the transistor and also provides the collector contact area. A shallow emitter is next implanted and/or diffused into the device, followed by contacts and metallization. Typical parameters for the CDI process include a buried-layer sheet resistance of 15 to 30 ohms per square, a 2-μm-thick, 0.25-ohm-cm epitaxial layer, and an emitter depth of less than 1 μm.

This process can produce high-performance, narrow-base transistors with minimum r_C but with relatively large collector-base and collector-substrate capacitances. It has for the most part been replaced with advanced oxide isolated structures which also minimize these capacitances, although at a cost of considerable process complexity.

10.8.2   V-Groove Isolation

V-groove isolation[8–9] attempts to produce small-geometry, high-frequency, bipolar transistors by eliminating the capacitance of junction isolation, as shown in Fig. 10.13. Anisotropic etching produces V-grooves which separate the transistors. ⟨100⟩ silicon must be used in this process to permit formation of the grooves.

The process looks very similar to the SBC process through diffusion of the p-type base region into the n-type epitaxial layer. The V-grooves are etched completely through the epitaxial layer, thereby achieving device isolation. An oxide-nitride sandwich is used to passivate the surface of the structure prior to formation of the n^+ emitter and collector contact regions. For shallow structures, a multilayer metallurgy is used to make contacts to the transistor, and several levels of metallization are used for circuit interconnection.

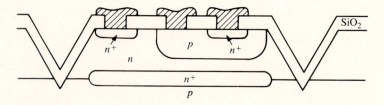

Fig. 10.13   A bipolar transistor formed using V-groove isolation.[8–9]

10.8.3 Dielectric Isolation

Figure 10.14 shows the steps used to achieve dielectrically isolated bipolar transistors.[10] Deep V-grooves are etched in the surface of $\langle 100 \rangle$-oriented silicon. A nonselective n^+ diffusion is performed, and a silicon dioxide layer is grown on the surface. A thick layer of polycrystalline silicon is then deposited on the surface of the wafer. Silicon is removed from the back surface of the wafer by lapping until the silicon dioxide in the V-grooves is exposed, as indicated by the dotted line in Fig. 10.14b. The surface is then mechanically and chemically polished. The wafer is turned over, yielding islands of silicon which are completely isolated from each other by the silicon dioxide dielectric layer. Standard processing is then used to form bipolar transistors.

This process is expensive, but an important variation permits formation of complementary, vertical *npn* and *pnp* transistors. It is used in fabrication of high-performance analog circuits. In addition, the structures are very tolerant to radiation and so are used in military applications.

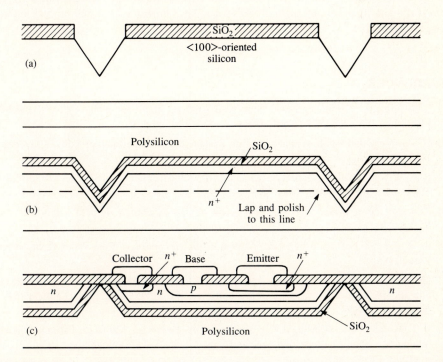

Fig. 10.14   Several steps in the dielectric isolation process.[10] (a) V-grooves anisotropically etched in the silicon substrate; (b) the structure following n^+ diffusion and oxidation; the wafer is turned over and lapped back and polished to the dotted line; (c) bipolar transistors are then fabricated in the isolated islands of silicon.

10.9 SUMMARY

The *standard buried collector* (SBC) process is widely used throughout the IC industry for analog and power circuit applications. More recently developed digital bipolar technologies have benefited greatly from process advances originally developed for use in MOS dynamic RAMs. These include oxide isolation, the use of polysilicon, and the introduction of ion implantation.

Bipolar technologies for analog applications are typically designed to yield current gains of several hundred with breakdown voltages of up to 50 V. The resulting devices have cutoff frequencies of less than 500 MHz. Devices for digital applications can operate with much lower current gains and supply voltages. These factors permit device designs with cutoff frequencies exceeding 5 GHz.

Collector-diffusion isolation, V-groove isolation, and dielectric isolation have all been developed as alternatives to the junction-isolated SBC process. These processes demonstrate the freedom that the process designer has in trying to achieve new and improved isolation techniques. However, the SBC process and oxide-isolated bipolar process remain the dominant technologies for today's analog and digital bipolar applications, respectively.

REFERENCES

[1] J. G. Fossum, "Computer-Aided Numerical Analysis of Silicon Solar Cells," Solid-State Electronics, *19,* 269–277 (April, 1976).

[2] H. Lawrence and R. M. Warner, Jr., Bell System Technical Journal, *39,* 389–403 (March, 1960).

[3] P. R. Wilson, "The Emitter-Base Breakdown Voltage of Planar Transistors," Solid-State Electronics, *17,* 465–467 (May, 1974).

[4] R. A. Colclaser, *Microelectronics Processing and Device Design,* John Wiley & Sons, New York, 1980.

[5] S. M. Sze, *Semiconductor Devices —Physics and Technology,* John Wiley & Sons, New York, 1985.

[6] M. Vora, Y. L. Ho, S. Bhamre, F. Chien, G. Bakker, H. Hingarh, and C. Schmitz, "A Sub-100 Picosecond Bipolar ECL Technology, IEEE IEDM Technical Digest, p. 34–37 (December, 1985).

[7] B. T. Murphy, V. J. Glinski, P. A. Gary, and R. A. Pederson, "Collector Diffusion Isolated Integrated Circuits," Proceedings of the IEEE, *57,* 1523–1527 (September, 1969).

[8] J. Mudge and K. Taft, "V-ATE Memory Scores a New High in Combining Speed and Bit Density," Electronics, *45,* 65–69 (July 17, 1972).

[9] T. J. Sanders, W. R. Morcom, and C. S. Kim, "An Improved Dielectric-Junction Combination Isolation Technique for Integrated Circuits," IEEE IEDM Technical Digest, p. 38–40 (December, 1973).

[10] J. L. Davidson and D. R. Mason, Method of Etching Silicon Crystals, U.S. Patent #3,728,179, issued April 17, 1973.

[11] C. C. Allen, L. H. Clevenger, and D. C. Gupta, "A Point Contact Method of Evaluating Epitaxial Layer Resistivity," Journal of the Electrochemical Society, 113, 508–510 (May, 1966).

[12] H. F. Wolf, Silicon Semiconductor Data, Pergamon Press, Oxford, 1969.

PROBLEMS

10.1 Evaluate the Gummel number expressions for a uniformly doped transistor with impurity concentrations of N_E and N_B in the emitter and base, respectively. The effective width of the emitter is L_E, and W_B is the basewidth. Compare your result with eq. (2.46) in Volume III of this Series. What is the current gain for a device with $N_E = 10^{20}/cm^3$, $N_B = 10^{18}/cm^3$, $W_B = 4$ μm, $L_E = 20$ μm, $L_B = 50$ μm, $D_B = 20$ cm^2/sec, and $D_E = 5$ cm^2/sec? Assume $\eta = 1$.

10.2 Using eq. (10.1), estimate the current gain of the transistor with the impurity profiles given in Fig. 10.2. Assume $W_B \ll L_B$, and $D_B = 20$ cm^2/sec. Use $N_E/D_E = 5 \times 10^{13}$ sec/cm^4, $x_{JE} = 1.5$ μm, $x_{JC} = 4$ μm.

10.3 What is the maximum collector-base breakdown voltage of a transistor with $X_{BL} - X_{BC} = 5$ μm? What range of epitaxial layer dopings may be used to achieve this breakdown voltage?

10.4 A Zener reference diode is often formed using breakdown of the emitter-base junction.

(a) What base surface concentration is required to produce a breakdown voltage of 6 V for a 1-μm-deep junction?

(b) If the base surface concentration is too small by a factor of two, what is the actual breakdown voltage of the diode?

10.5 Calculate the collector-base depletion-layer width for the transistor of Example 10.3 using the expression for a one-sided step junction given in eq. (9.3). How does this compare with the width derived from Fig. 10.4?

10.6 The effective Gummel number in the emitter is substantially reduced from that calculated from the profile by bandgap narrowing in the emitter. Calculate G_E using the expressions in Chapter 4 for an As emitter with a sheet resistance of 10 ohms per square. Compare G_E to the values stated in the text.

10.7 In Section 10.6.3 the importance of correct positioning of the n^+ collector contact diffusion was discussed. To illustrate this point, three simple npn transistors are fabricated in an n-type substrate using the structure drawn in Fig. P10.7. An n^+ collector ring is used to reduce the collector series resistance R_C. Three different spacings, 0 μm, 3 μm, and 5 μm, are used between the edge of the n^+ ring and the edge of the base diffusion. Explain why the breakdown voltage will be different for these three cases, and estimate the collector-base breakdown voltage for the three devices. Assume a base surface concentration of $10^{18}/cm^3$ and a junction depth of 5 μm.

10.8 Surface conditions can degrade the breakdown voltage of Zener diodes. Subsurface breakdown can be achieved using an ion-implanted process with the profile shown in Fig. P10.8. Estimate the breakdown voltage of this diode.

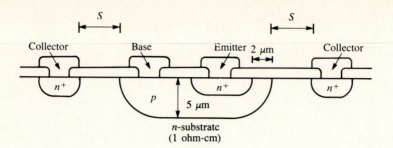

Fig. P10.7

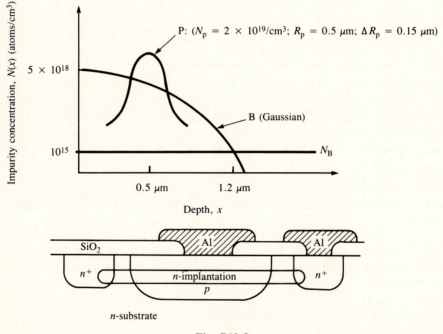

P: ($N_p = 2 \times 10^{19}/cm^3$; $R_p = 0.5~\mu m$; $\Delta R_p = 0.15~\mu m$)

Fig. P10.8

10.9 A simple lateral *pnp* structure is shown in Fig. P10.9. The current gain of this transistor is collection-limited and does not obey eq. (10.1). Assume that the emitter injects current uniformly in all directions and that the collector collects all the current coming its way.

(a) Under these assumptions, what is the value of the common base current gain $\alpha = I_C/I_E$? What is the common emitter current gain β?

(b) Derive an expression for β as a function of the length and width of the device. For a given area, what relationship between the length and width maximizes the gain?

(c) What geometry would be used to optimize the current gain?

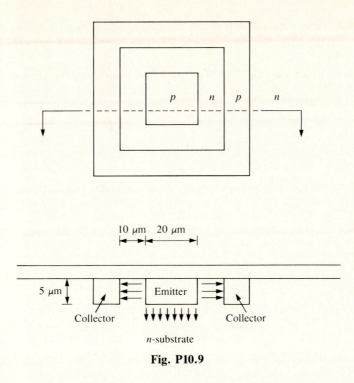

Fig. P10.9

10.10 It was noted that the CDI process is used mainly for digital applications. What characteristics of the structure make this true?

10.11 (a) How many masks are required for the CDI process?

(b) Design a good mask-alignment sequence for this process.

10.12 List the mask steps required for the oxide-isolated bipolar transistor of Fig. 10.11. Which are noncritical alignment steps?

10.13 Determine a reasonable diffusion schedule for the isolation diffusion of a junction-isolated structure with a 15-μm-thick epitaxial layer with the geometry of Fig. P10.13.

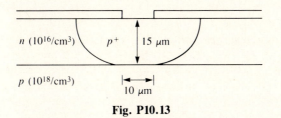

Fig. P10.13

(a) Assume that the width of the isolation at the bottom is to be 10 μm, that there is no up-diffusion from substrate, and that lateral diffusion equals vertical diffusion.

(b) Modify your diffusion time in part (a) to account for up-diffusion of boron from the substrate. Assume that the substrate represents an infinite supply of boron impurities with a constant concentration of $10^{18}/cm^3$.

10.14 The distance $X_{BL} - X_{BC} = 5$ μm in a certain bipolar SBC process with an epitaxial layer doping of $10^{15}/cm^3$. Use the one-sided step-junction formula, eq. (9.3), to estimate the punch-through voltage for this transistor. Compare with Fig. 10.6.

10.15 V-grooves in $\langle 100 \rangle$ silicon make angles of 54.7° with the surface. What is the minimum isolation groove width at the surface if the epitaxial layer is 5 μm thick and we require a 1-μm minimum isolation width at the bottom of the groove? Does this seem competitive with other isolation processes? Which ones?

10.16 A CDI process uses a 0.25-ohm-cm epitaxial base layer and a 5-ohm-cm substrate. Estimate the breakdown voltages of the emitter-base and collector-base junctions. The emitter junction depth is 1 μm, and the expitaxial layer thickness is 2 μm.

Answers to Selected Problems

CHAPTER 1

1.1   Approximately 256
1.2   75 mm, 130 dice, $8.79; 150 mm, 600 dice, $1.90
1.3   10,000; 160,000; 1,000,000

CHAPTER 2

2.1   92.3%

CHAPTER 3

3.2   27 min, 63 min, 102 min
3.4   0.07 μm, 0.4 μm
3.7   3.55 hr, orange

CHAPTER 4

4.1   5.8 μm, 5.3 μm, 47 and 60 ohms/square
4.3   5.14 hr, 83 ohms/square, $6.7 \times 10^{14}/cm^2$
4.7   10.7 squares, approximately 5 squares, 710 and 330 ohms
4.9   23%
4.13   5 min
4.14   100 ppm

CHAPTER 5

5.1 $3.33 \times 10^{18}/cm^3$, $1.1 \times 10^{13}/cm^2$, 0.40 μm
5.3 $5.06 \times 10^{14}/cm^2$, 10.1 hr
5.5   0.14 μm
5.7 $2 \times 10^{18}/cm^2$, 43.6 mA, 218 KW, probably melt the wafer
5.9   33 min

CHAPTER 6

6.1   120 μsec

6.3 $4.2 \times 10^{-7}/cm^3$

6.7   0.14 μm/min, 0.02 μm/min, 1245 °C, 0.20 μm/min

6.10   Approximately 7100 wafers

CHAPTER 7

7.1   0.032 ohms/square, 1.6 ohms, 0.175 pF, 0.28 psec

7.3   Boron: 7.8 ohms/square

7.4   0.08 μm

7.6   125, 2.53×10^{26}

7.8   20 mA

CHAPTER 8

8.2 1.3×10^{20} years/die, 1.3×10^{22} years/wafer

8.4   Depends on placement of die, best case 2/26 or 7.7%, worst case 0%; approximately 3.1

8.5 $9.66, $8.35 — cheaper!

8.7   Old process: Cost = wafer cost/161; New process: Cost = wafer cost/153; slightly more expensive; $D_0 = 9.4/cm^2$; yes, since we should move down learning curve on new process; 12.2 mm^2

8.9   There is an infinite number of solutions; for example, 14.8 mm^2 and 17.8 mm^2 give a die cost of $1 in each process.

CHAPTER 9

9.2 2×3.6 μm = 7.2 μm

9.3   14 volts

9.5   NMOS — 0.045 volts, $5.1 \times 10^{11}/cm^2$;
PMOS — -1.87 volts, $4.7 \times 10^{11}/cm^2$

9.6 $1.75 \times 10^{12}/cm^2$

CHAPTER 10

10.3   Approximately 70 volts

10.5   1.1 μm

10.8   3 volts, 30 volts, 50 volts

10.9   Approximately 3 volts

10.10   0.5, 1.0, 2d(L + W)/LW where d is the diffusion depth; square with L = W and beta = 4d/L, circular

Index

PHYSICAL CONSTANTS

Symbol	Name	Value
q	Magnitude of electronic charge	1.602×10^{-19} C
m_0	Electron rest mass	9.109×10^{-31} kg
m_p	Proton rest mass	1.673×10^{-27} kg
c	Speed of light in vacuum	2.998×10^8 m/s
ε_0	Permittivity of vacuum	8.854×10^{-12} F/m
k	Boltzmann's constant	1.381×10^{-23} J/K 8.617×10^{-5} eV/K
h	Planck's constant	6.625×10^{-34} Js 4.135×10^{-15} eVs
A_0	Avogadro number	6.022×10^{26} molecules/kg-mole
kT	Thermal energy	0.02586 eV ($T = 27$ °C) 0.02526 eV ($T = 20$ °C)
E_g	Bandgap of silicon at 300K	1.12 eV
K_s	Relative permittivity of silicon	11.7
K_0	Relative permittivity of silicon dioxide	3.9
n_i	Intrinsic carrier density in silicon at 300K	$10^{10}/cm^3$

CONVERSION FACTORS

$1 \text{ Å} = 10^{-8}$ cm

$\quad\quad = 10^{-10}$ m

$1 \text{ } \mu m = 10^{-4}$ cm

$\quad\quad = 10^{-6}$ m

$1 \text{ mil} = 10^{-3}$ in.

$\quad\quad = 25.4 \text{ } \mu m$

$1 \text{ mil}^2 = 645.2 \text{ } \mu m^2$

$\quad\quad = 6.45 \times 10^{-6} \text{ cm}^2$

$1 \text{ eV} = 1.602 \times 10^{-19}$ J

$\lambda(\mu m) = 1.24/E(eV)$